T0202300

Fabio Pistolesi

Evolution from BCS Superconductivity to Bose-Einstein Condensation and Infrared Behavior of the Bosonic Limit

TESI DI PERFEZIONAMENTO

SCUOLA NORMALE SUPERIORE
1998

Tesi di perfezionamento in Fisica sostenuta il 30 gennaio 1997

COMMISSIONE GIUDICATRICE

To my parents

Contents

Introduction

A few words are necessary to explain the title and the structure of this Thesis.

The main aim of the work was initially to study the possible relevance of the cross-over from BCS superconductivity to Bose-Einstein condensation to explain some unifying features found in experiments on high temperature superconductors. The first Part of the Thesis is in fact dedicated to this subject. But during our work we have gradually recognized that many of the technical and physical difficulties that we had to face in the crossover problem were essentially present already in the purely Bose system.

We passed then to study more closely the Bose system itself. We found that, notwithstanding the huge number of papers published on the subject, there were still many fundamental unanswered questions. In particular, the fact that perturbation theory was plagued by infrared divergences leads to puzzling inconsistencies. The most evident of them is the vanishing of the exact anomalous self energy, in contrast with the finite result of lowest orders calculations. This result questions the validity of Boson perturbation theory, which is essentially the same perturbation theory that we were using to describe the Fermi system.

We realized at this point that to gain a full control on the crossover phenomenon we needed a more complete understanding of the Bose system beforehand. In this second phase of the work we started the collaboration with Prof. C. Castellani and Prof. C. Di Castro. So the second Part of the Thesis is dedicated to the Renormalization Group approach for the zero-temperature interacting Bosons; in this way we have healed the divergences of the theory and found the exact infrared behavior of the Bose system.

Coming back to the Fermi problem is, at this point, beyond the scope of this Thesis, but it can be a promising line for future research.

We refer the reader to the Introductions of the two Parts for more details and for a complete list of references. Results of the first Part are already published [1–4], a short account of the results of the second Part has been given [5], and a longer version is in preparation [6].

Part I

Crossover from BCS to BEC

Chapter 1

Introduction to Part I

There has been recently renewed interest in the crossover from BCS superconductivity to Bose-Einstein (BE) condensation [7–18], following the discovery
of the high-temperature superconductors [19]. In particular, the observation
that these (as well as other "exotic") superconductors have considerably (i.e.,
$10^3 - 10^4$ times) shorter *coherence length* than conventional superconductors
has prompted the suggestion that proper description of superconductivity in
these materials might require an *intermediate* approach between the two limits represented by BCS theory and BE condensation [20]. In this context, it
appears especially relevant to assess how the coherence length (which can be
determined experimentally from the spatial fluctuations of the order parameter and which we shall consistently refer to as ξ_{phase} in the following) crosses
over between these two limits. Purpose of this Part is to provide a detailed
description of this crossover [21].

Evolution from weak- to strong-coupling superconductivity has been addressed a few years ago by Nozières and Schmitt-Rink [22] (hereafter referred
to as NSR) after the pioneering works by Keldish and Kopaev [23], Keldish
and Kozlov [24], Eagles [25] and Leggett [26]. NSR follow this evolution by
increasing the coupling strength of an effective fermionic attractive potential,
and conclude that the evolution is "smooth". The inclusion of fluctuations beyond mean field considered by NSR through the ladder approximation for the
pairing susceptibility, however, has posed problems of physical consistency [27],
owing to the fact that the ladder approximation is not "conserving" [28]. This
shortcoming has been later overcome by Haussmann [29] who considered a fully
"conserving" diagrammatic approach to describe the interacting Fermi system
in the superconducting phase, whereby each single-particle Green's function is
self-consistently determined. It turns out that keeping the full self-consistency
is most important in the intermediate (crossover) region of interest, in order
to account correctly for the mixture of fermionic and bosonic degrees of freedom [29].

The approaches of Refs. 27 and 29 (as well as the related work of Refs. 7–14) rely on an approximation scheme (i.e., BCS mean field plus fluctuations) which is well-established in the weak-coupling limit. The fact that this procedure results into a sensible strong-coupling limit (i.e., the noninteracting Bose gas of Ref. 27 or the weakly interacting Bose gas of Ref. 29) can be related to the structure of the BCS wave function, which has built in the BE condensation has a limiting case [30]. There is, however, *a priori* no guarantee that the results in the strong-coupling limit would always provide a satisfactory description of the limiting system of interacting bosons.

For these reasons, we prefer to approach the bosonization process *in reverse*, that is, by setting up first a reliable approximation for the bosonic system and then determining how the bosonization procedure of the original fermionic system maps that approximation back onto a description of the weak-coupling limit. In this way, we can focus directly on improving the description of the bosonic limit, which is admittedly more difficult to deal with than the opposite weak-coupling limit, where the BCS approximation is expected to be invariably recovered as the fundamental starting point.

Focusing directly on the bosonic limit, however, leads us to confront with a long-standing problem in the theory of interacting bosons. It has, in fact, long been known that conventional many-body (diagrammatic) methods for an interacting *condensed* Bose system can be quite generally organized within approximation schemes that are consistent *either* with conservation laws ("Φ-derivable" approximations) *or* with the absence of a gap in the elementary excitations spectrum ("gapless" approximations) [31]. This difficulty does not appear in the corresponding scheme for self-consistent "Φ-derivable" (conserving) approximations for fermionic systems (*even* in the superconducting phase) [32]. For these reasons, when dealing with condensed bosonic systems one prefers to abandon self-consistent schemes and resorts instead to approximation procedures whereby diagrams are selected in terms of an *external* small parameter (like the reduced density) [33].

An approach formally alternative to conventional diagrammatic methods to set up a modified perturbation theory for a superfluid Bose system is the functional-integral method with the associated "loop" expansion, which allows for a unified description of superfluidity and superconductivity in terms of collective variables [34,35]. By this method, bosonic-like collective variables are introduced at the outset in the description of the fermionic superconducting system of interest via a Hubbard-Stratonovich transformation, in terms of which a mean-field approximation and the associated fluctuation corrections can be defined. Specifically, the mean-field approximation recovers the NSR results obtained at zero temperature, while systematic inclusion of fluctuation corrections by the loop expansion enables one to overcome the problems of physical consistency mentioned above for the NSR results [36]. In this context, it is worth mentioning the recent work by Traven [37] who considered the interaction between pair fluctuations (which are ignored by the standard Gaussian approximation) and demonstrated that it removes the pathological behavior of the thermodynamic functions obtained within the Gaussian approximation in two dimensions, thus stabilizing the low-temperature superfluid phase. We have also mention in this context that the loop expansion associated with the functional integral can be formally mapped [38] in the bosonic limit onto the low-density expansion (which is conventionally used to select the relevant diagrammatic structure for the *dilute* Bose gas [33]). Keeping all terms up to a given order in the expansion parameter further guarantees that conservation laws *and* Ward identities are satisfied up to the same order. It is in this sense that the problems originating from the "gapless" and "Φ-derivable" approximations are overcome by the "loop" expansion. In the following, we shall apply the functional-integral method at the one-loop (i.e., the next-to-significant) order to the problem of the crossover between BCS and BE, with the same model Hamiltonian adopted by NSR.

Returning, specifically, to the calculation of the phase coherence length ξ_{phase} at zero temperature, we will show that the one-loop calculation leads to a consistent picture for the crossover of this physical quantity, which varies from the Pippard coherence length ξ_0 in the weak-coupling limit to the known result $(4m_B\mu_B)^{-1/2}$ for a dilute Bose gas (with mass m_B and chemical potential μ_B) in the strong-coupling limit [33]. These results will be contrasted with the (mean-field) calculation of the coherence length ξ_{pair} for two-electron correlation reported previously [39], (and discussed in 2) which ranges instead from ξ_0 to the bound-state radius r_0 in the two limits. In Ref. 39 it was also concluded that (i) ξ_{pair} (through the dimensionless parameter $k_F\xi_{pair}$) is the relevant variable to *follow* the crossover from BCS to BE (and thus it does not serve to identify merely the two extreme BCS ($k_F\xi_{pair} \gg 1$) and BE ($k_F\xi_{pair} \ll 1$) limits), and (ii) this crossover occurs in practice in a limited range of the variable $k_F\xi_{pair}$ (beginning at $k_F\xi_{pair} \simeq 10$ on the BCS side). The calculation of ξ_{phase} reported in 3 confirms this result, because we will find that $\xi_{phase} \simeq \xi_{pair}$ *down to* $k_F\xi_{pair} \simeq 10$, with the two lengths starting to differentiate for smaller values of $k_F\xi_{pair}$ [40]. One can thus associate a *single* characteristic length to a BCS-type superconductor, which is generically identified (even for relatively strong coupling) by the existence of a well-defined Fermi surface. In the bosonic limit, we will find instead that $\xi_{phase} \gg \xi_{pair}$, as expected, since the "size" of a single boson is by no means related to the range of the fluctuations of the order parameter [41]. We shall further show in this limit that ξ_{pair} is associated with the range of the *residual* boson-boson interaction, by mapping the original fermionic system onto an effective system of interacting bosons. In this way, a sensible and consistent description of the bosonization process results from our one-loop calculation, at least in the zero-temperature limit we are considering.

The plan of Part I is the following. In Chapter 2 we consider the phenomenological use of ξ_{pair} to rationalize Uemura plot. In Chapter 3 we set up the calculation of ξ_{phase} at the one-loop order, by relying on a functional-integral representation of the correlation functions for a fermionic system interacting through an effective potential of the type introduced by NSR. We provide also analytic expressions of ξ_{phase} in the weak- and strong-coupling limits. In Chapter 4 we consider specifically the strong-coupling limit and perform a mapping of the effective action of the original fermionic system onto the corresponding action of a truly bosonic system, by exploiting features of the collective bosonic-like variables introduced in Chapter 3 via a Hubbard-Stratonovich transformation. In Chapter 5 we present numerical results for ξ_{phase} (in three and lower dimensions) over the whole range of coupling, and especially across the *narrow* region of the variable $k_F\xi_{pair}$ where the actual crossover from BCS to BE takes place. Chapter 6 gives our conclusions. Details of the calculations as well as related additional material are given in the Appendices. In particular, Appendix A obtains the shift of the order parameter which is required to make the NSR approach fully consistent at the one-loop level in the condensed phase (or in two dimensions [37]). In Appendix D the "universal" curve, obtained previously in Ref. 39 for the chemical potential versus $k_F\xi_{pair}$ using the NSR separable interaction, is discussed further in the context of the (three-dimensional) negative-U Hubbard model and of the analytic two-dimensional solution of Ref. 8.

Chapter 2

The Uemura Plot

2.1 Experimental Results

Y.J. Uemura *et al.* [42] have recently proposed to distinguish a class of "exotic" superconductors (including cuprate, bismuthate, organic, Chevrel-phase, heavy-fermions, and fullerene systems) from the more conventional superconductors (like Nb) by the value of the ratio T_c/T_F between the experimental superconducting temperature and the (effective) Fermi temperature, which turns out to be about one hundred times larger for the "exotic" than for the conventional superconductors. Y.J. Uemura *et al.* have also suggested that this substantial difference might be an indication that the "exotic" superconductors are in some sense intermediate between conventional BCS superconductors and Bose-condensed systems.

In order to establish a possible connection between the physical consequences of this suggestion and the experimental relation between T_c and T_F, one should first try to figure out the appropriate variable to determine the crossover between Cooper-pair-based (BCS) superconductivity and Bose-Einstein (BE) condensation. One could then attempt to use that variable as a sort of "normal" coordinate in the plot constructed by Y.J. Uemura *et al.* (hereafter referred to simply as the Uemura plot), by figuring out a phenomenological relation (independent of the underlying superconducting mechanism) between that variable and the ratio T_c/T_F, thus encompassing the difference between the "exotic" and conventional superconductors. Completion of this program might further turn out to be useful both experimentally (by suggesting possibly in which direction one should move in the Uemura plot to improve T_c and theoretically (by giving hints on the underlying dynamics related to high-T_c superconductivity. In this chapter we shall propose how to meet this program by a rather general argument on the stability of the Cooper-pair-based superconductivity

2.2 The NSR Model

Evolution from weak to strong coupling superconductivity has been addressed a few years ago by Nozières and Schmitt-Rink [22] following the pioneering work by Legget [26]. Central to this work is the well-known argument [43] that the BCS wave function has built in the Bose-Einstein (BE) condensation as a limiting case, since it reduces to the BE condensate wave function when the (average) occupation numbers $\langle n_{\mathbf{k}\sigma} \rangle$ can be neglected with respect to unity for *all* wave vectors $\mathbf{k}$ (and for both spin projections σ). NSR follow the evolution from Cooper-pair-based superconductivity to BE condensation through the increase of the coupling strength associated to an effective fermionic attractive potential, and conclude that the evolution is "smooth". Although this result is appealing from a theoretical point of view, it does not allow for a direct comparison with the Uemura plot since the coupling strength of the effective fermionic attraction is not a quantity that could be realistically inferred from experiments. Besides, associating an effective coupling strength to a given class of superconductors would unavoidably require one to face at the outset the problem of the mechanism responsible for superconductivity [44], which need not be actually necessary to unravel the kind of message conveyed by the Uemura plot.

We need thus to figure out a more significant variable than the coupling strength to follow the evolution from BCS superconductivity to BE condensation. The new variable should be selected according the following criteria: (*i*) the evolution from BCS to BE should turn out to be as much as possible *universal*, i.e., independent of the details of the interaction potential and of the single-particle density of states; (*ii*) when using the new variable to rationalize the Uemura plot, it should be possible to subject that variable to an *independent* experimental check [45]. We shall illustrate in the following why we propose to identify the product $k_F \xi$ as the desired variable.

Following NSR, we introduce the model fermionic Hamiltonian

$$
\begin{aligned}
H &= \sum_{\mathbf{k}\sigma} \epsilon_{\mathbf{k}} c_{\mathbf{k}\sigma}^{\dagger} c_{\mathbf{k}\sigma} \\
&+ \sum_{\mathbf{k},\mathbf{k}',\mathbf{q}'} V_{\mathbf{k},\mathbf{k}'} c_{\mathbf{k}\mathbf{q}/2\uparrow}^{\dagger} c_{-\mathbf{k}\mathbf{q}/2\downarrow}^{\dagger} c_{-\mathbf{k}'\mathbf{q}/2\downarrow} c_{\mathbf{k}'\mathbf{q}/2\uparrow} ,
\end{aligned} \tag{2.1}
$$

where $c_{\mathbf{k}\sigma}$ is the destruction operator for fermions with wave vector $\mathbf{k}$ and spin σ, $\epsilon_{\mathbf{k}}$ is a single-particle (or quasi-particle) dispersion relation, and $V_{\mathbf{k}\mathbf{k}'}$ is an "effective" fermionic attraction. In its simplest version, $\epsilon_{\mathbf{k}} = \mathbf{k}^2/2/m^* - \mu$ where m^* is an effective (quasi)particle mass and μ is the chemical potential [46]. Although the use of the Hamiltonian (2.1) has obvious shortcomings, by resting on a continuum model (where no effect of the lattice structure is included) and by disregarding dynamical effects, we believe that it is sufficient for our purposes.

The variational procedure with the usual BCS trial wave function

$$|\Phi\rangle = \prod_{\mathbf{k}} \left(u_{\mathbf{k}} + v_{\mathbf{k}} c_{\mathbf{k}\uparrow}^{\dagger} c_{-\mathbf{k}\downarrow}^{\dagger} \right) |0\rangle \tag{2.2}$$

leads to the two familiar *coupled* equations

$$2\epsilon_{\mathbf{k}} \phi_{\mathbf{k}} + (1 - 2v_{\mathbf{k}}^2) \sum_{\mathbf{k}'} V_{\mathbf{k}\mathbf{k}'} \phi_{\mathbf{k}'} = 0 \tag{2.3}$$

$$.n = \frac{1}{\Omega} \sum_{\mathbf{k}} \left(1 - \frac{\epsilon_{\mathbf{k}}}{E_{\mathbf{k}}} \right) \tag{2.4}$$

where n is the particle density, Ω the quantization volume, and $\phi_{\mathbf{k}} = 2u_{\mathbf{k}} v_{\mathbf{k}} = \Delta_{\mathbf{k}} E_{\mathbf{k}}$ with $\Delta_{\mathbf{k}} - \sum_{\mathbf{k}'} u_{\mathbf{k}'} v_{\mathbf{k}'}$ and $E_{\mathbf{k}} \sqrt{E_{\mathbf{k}}^2 + \Delta_{\mathbf{k}}^2}$ [47]. Notice that, provided that $v_{\mathbf{k}}^2 \ll 1$ for *all* $\mathbf{k}$, Eq.(2.3) reduces to the Schrödinger equation for the relative motion of two particles with equal mass m^*, interacting via $V_{\mathbf{k}\mathbf{k}'}$ and with eigenvalue 2μ. In this limit bosonization of bound-electron pairs is fully achieved.

Solution of Eq. (2.3) and (2.4) gets considerably simplified by considering a separable potential. NSR choose

$$V_{\mathbf{k}\mathbf{k}'} = V w_{\mathbf{k}} w_{\mathbf{k}'}, \qquad w_{\mathbf{k}} = \frac{1}{\sqrt{1 + (\mathbf{k}/k_0)^2}}, \tag{2.5}$$

where the strength $V(< 0)$ and the characteristic wave vector k_0 are the parameters of the interaction. With this potential, $\Delta_{\mathbf{k}} = \Delta_o w_{\mathbf{k}}$ and the associated two-body eigenvalue problem has (in three dimensions) the only eigenvalue $2\mu = -\epsilon_o = -(k_0^2/m^*)(G - 1)^2$ for $G > 1$, where $G = -V\Omega m^* k_0/4\pi > 0$ is the dimensionless coupling constant, while the bound-state radius has the asymptotic behavior $r_0 \sim k_0^{-1} G^{-1/2}$ for $G \to \infty$ [48]. In agreement with the results by NSR, we find that the solutions (Δ_o, μ) of Eqs.(2.3) and (2.4) evolve smoothly as functions of G *for given* k_0, although $(\Delta_o\ \mu)$ depend strongly on k_0 for given G [49]. No connection with the Uemura plot is evidently possible at this level.

2.3   The New Variable

A suggestion to figure out a more significant variable than the coupling strength comes from the observation that "exotic" superconductors of the Uemura plot have a considerable shorter *coherence length* ξ ($\sim 20 - 50$Å) than the conventional superconductors (for which $\xi \sim 10^3 - 10^4$Å). Theoretically, ξ can be obtained from the pair correlation function with opposite spins

$$g(\mathbf{r}) = \frac{1}{n^2} \left| \left\langle \Phi \left| \Psi_{\uparrow}^{\dagger}(\mathbf{r}) \Psi_{\downarrow}^{\dagger}(0) \right| \Phi \right\rangle \right|^2 \tag{2.6}$$

where $\Psi_\sigma(\mathbf{r})$ is the fermion field operator, by identifying

$$\xi^2 = \frac{\int d\mathbf{r}g(\mathbf{r})\mathbf{r}^2}{\int d\mathbf{r}g(\mathbf{r})} = \frac{\sum_{\mathbf{k}}|\nabla_{\mathbf{k}}\phi_{\mathbf{k}}|^2}{\sum_{\mathbf{k}}|\phi_{\mathbf{k}}|^2} \qquad (2.7)$$

Our definition (2.7) in the weak-coupling limit correctly reduces to the Pippard coherence length $\xi_0 = (d\epsilon_{\mathbf{k}}dk)_{k_F}/\pi\Delta_{k_F}$ (actually, $\xi \to \xi_0\pi/2\sqrt{2} = 1.11\xi_0$ when $G \ll 1$) and to the bound-state radius r_0 in the strong-coupling limit. It turns out that the behavior of ξ versus G is also "smooth", although it strongly depends on the parameter k_0 of the interaction. A reasonable attempt to eliminate the coupling constant G from further considerations is then to replace it by a dimensionless parameter containing ξ. Since k_F^{-1} is the only other independent physical length scale in the problem, we replace the original pair of variables (G, k_0) by the alternative pair (ξ, k_0) and study the crossover from BCS superconductivity to BE condensation as a function of ξ for given k_0 [50].

In Fig. 2.1 we report the chemical potential μ versus ξ for a wide range of values of the reduced density $n/k_0^3 \ (= 10^\alpha$ with $\alpha = -5, -4, \ldots, +4)$. Positive values of μ have been normalized by the Fermi energy $\epsilon_F \ (= k_F^2/2m^*$, by our definition), while negative values of μ have been normalized by half of the eigenvalue ϵ_o of the associated two-body problem. The two curves for the reported extreme values of n/k_0^3 act as limiting (accumulation) curves for all practical purposes. The striking feature of Fig. 2.1 is that, when expressed in terms of ξ the behavior of the chemical potential becomes *universal* (i.e., independent of the parameter k_0 of the interaction potential), but possibly for an "intermediate" range $\pi^{-1} \lesssim \xi \lesssim 2\pi$ where the normalization values ϵ_F and ϵ_0 (depending of the sign of μ) loose actually their meaning [51]. This remarkable universal behavior of μ versus ξ strongly suggests that ξ *is indeed the appropriate variable* to follow the evolution from BCS superconductivity to BE condensation.

Notice in addition from Fig. 2.1 that μ gets pinned to (about) the normal-state value ϵ_F when $\xi \gtrsim 2\pi$, and that μ drops *rather abruptly* from ϵ_F at $\xi \simeq 2\pi$. In other words, Fig. 2.1 shows that, when the coherence length ξ equals the Fermi wavelength $\lambda_F = 2\pi/k_F$ of the electrons, the system becomes unstable against bosonization and the Fermi surface is wiped out. We expect that the instability of the Cooper-pair-based superconductivity when $\xi \simeq 2\pi$, inasmuch as it is consequence of a genuine quantum-mechanical effect, should actually persist beyond the limits of validity of the procedure we have followed to establish it. In this sense, the stability criterion $\xi \gtrsim 2\pi$ should be regarded the analog for the problem at hand of the Ioffe-Regel criterion for transport in disordered systems.

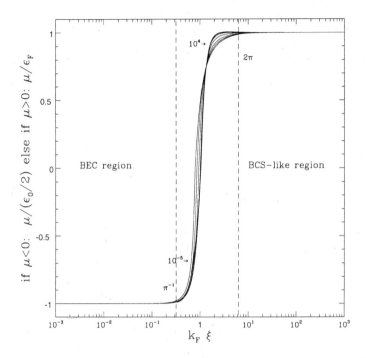

Figure 2.1: Chemical potential μ versus ξ (at zero temperature). The normalization of μ and the meaning of the different curves are explained in the text. The two limiting curves corresponding to the values 10^{-5} and 10^4 of the reduced density n/k_0^3 are indicated.

2.4   Interpretation of the Plot

It still remains to figure out how the Uemura plot for T_c versus T_F could be mapped out in terms of the variable ξ. To this end, we remark that the variable ξ_0 appears in the characteristic weak-limit BCS expression for T_c obtained from $k_B T_c / \Delta_{k_F} = e^\gamma / \pi$ (γ being Euler's constant) by eliminating Δ_{k_F} in favor of the Pippard coherence length ξ_0, namely

$$k_B / T_c = \frac{2e^\gamma}{\pi^2} \frac{\epsilon_F}{k_F \xi_0} \qquad (2.8)$$

k_B being Boltzmann's constant and $2e^\gamma/\pi^2 = 0.36$. The question naturally arises whether an expression like (2.8) could be used not only asymptotically in the weak-coupling limit, but also down to $\xi \simeq 2\pi$ whenever the concept of a Fermi surface is still preserved. To answer this question, we have followed again NSR and evaluated the thermodynamic potential within the ladder-approximation in the particle-particle channel for the normal state (i.e., for

F. Pistolesi

$T \geq T_c$). In this way, T_c and $\mu(T_c)$ have been determined versus G, or alternatively versus ξ for given k_0. The result is that the relation

$$k_B/T_c = 0.40 \frac{\epsilon_F}{\xi} \qquad (2.9)$$

holds universally for $\xi \gtrsim 2\pi$, independent of k_0 [52] (on the role of k_0 see also the discussion in 4.)

We can now envisage interpreting the Uemura plot in terms of the variable ξ by assuming Eq. (2.9) to hold phenomenologically (that is, irrespective of the assumptions used to derive it) and thus superimposing on the $\log_{10} T_c$ versus $\log_{10} T_F$ plot (with $T_F = \epsilon_F k_B$) the straight lines with ξ constant.

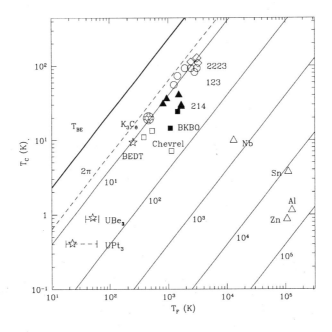

Figure 2.2: Uemura plot with superimposed lines of $\log_{10} T_c$ versus $\log_{10} T_F$ according to Eq. (2.9) of the text for the values 2π (broken line) and 10, 10^2, 10^3, 10^4, 10^5 (full lines) of ξ Experimental points are reproduced from Ref. [42].

The result is reported in Fig. 2.2 where the lines with $\xi = 10$, 10^2, 10^3, 10^4 and 10^5 are drawn (full lines) together with the "instability" line $\xi 2\pi$ (broken line). We notice the following features from Fig. 2.2

(i) a *linear trend* of $\log_{10} T_c$ versus $\log_{10} T_F$ results for any given value of ξ

(ii) the line with $\xi = 2\pi$ remarkably appears to be the natural *upper boundary* for the experimental data, suggesting that the systems near the boundary are close to a Fermi surface *instability*;

(*iii*) what distinguishes the "exotic" from the conventional superconductors in the Uemura plot is the smaller value of ξ associated to the former ones (apart from a possible difference in m^*);

(*iv*) high values of T_c results when ξ is "small" (~ 10) and m^* is not too large [53].

To obtain an independent check on whether our proposal to interpret the Uemura plot is correct, one should compare the values of ξ associated via Eq.(2.9) to the various samples in Uemura plot with *independently* measured values of ξ_{exp} (as obtained by critical magnetic field measurements) [54]. Preliminary checks with the experimental data available to us give indeed encouraging results for this comparison. A complete list of data on T_c/T_F, k_F and ξ_{exp} for all samples reported in the Uemura plot is, however, required to draw a definite conclusion about the success of the comparison.

2.5 Comments

Some final comments are in order. First, we remark that intrinsic to the procedure how the values of T_F are located in the Uemura plot is an effective angular averaging which washes out all non-spherical features of the Fermi surface. Otherwise, no meaningful connection between the Uemura plot and our expression (2.9) could even have been attempted. It may thus be possible that more specific criteria for instability toward bosonization, which would take into account the asymmetry of the Fermi surface (especially in reduced dimensionality), could result in smaller values (than 2π) for the product of the relevant values of k and of the coherence size. Second, we emphasize that in our approach we have not taken into account any lattice effect. It is likely the tendency toward bosonization may result into an instability of the coupled electron-lattice system (such as the onset of a charge-density wave or a structural modulation) that could actually overwhelm superconductivity This could be possibly the reason why no physical system appears to cross the "forbidden" boundary at $\xi \simeq 2\pi$ in the Uemura plot. A consequence of this sort of speculations would thus be that producing samples with even higher T_c would require one to move along the line $\xi \simeq 2\pi$ in the Uemura plot.

Chapter 3

Calculation of the $T = 0$ Phase Coherence Length for a Superconducting Fermionic System

In this Chapter we consider the calculation of the spatial fluctuations of the order parameter $\langle \psi_\uparrow(\mathbf{r})\psi_\downarrow(\mathbf{r}) \rangle$ for a superconducting fermionic system ($\psi_\sigma(\mathbf{r})$ being the fermionic field operator with spin projection σ), and follow their evolution as the fermionic system is driven toward its bosonic counterpart by increasing the (effective) attractive fermionic interaction. To this end (and for the reasons discussed in the Introduction), we shall rely on a functional-integral representation of the correlation functions and the associated loop expansion. Since some confusion has sometimes arisen in the literature between the inclusion of Gaussian fluctuations and the consistency of the loop expansion [13, 27], we shall discuss the latter in some detail in the following (see also Appendix A).

To identify the range of the spatial fluctuations of the superconducting order parameter across the evolution from BCS to BE, we find it convenient to introduce the bosonic-type operator

$$\varphi(\mathbf{R}) = \int d\rho \, \phi(2\rho)\psi_\uparrow(\mathbf{R} - \rho)\psi_\downarrow(\mathbf{R} + \rho) \tag{3.1}$$

where the (real) function $\phi(\rho)$ is assumed to be "localized" about $\rho = 0$. The thermal average $\langle \varphi(\mathbf{R}) \rangle$ can be then associated with the order parameter of the broken-symmetry phase. Since this order parameter is (in general) complex, we can further represent the operator (3.1) via its *longitudinal* and *transverse* components to the direction of broken symmetry [55]:

$$\varphi_\parallel(\mathbf{R}) \quad = \quad \frac{1}{2|\Delta|}\left(\Delta^* \varphi(\mathbf{R}) + \Delta \varphi^\dagger(\mathbf{R})\right) \tag{3.2}$$

$$\varphi_\perp(\mathbf{R}) \;=\; \frac{1}{2i|\Delta|}\left(\Delta^*\varphi(\mathbf{R}) - \Delta\varphi^\dagger(\mathbf{R})\right) \tag{3.3}$$

where we have set

$$\Delta = \langle\varphi(\mathbf{R})\rangle. \tag{3.4}$$

The relevant correlation functions for the operators (3) can then be defined as follows:

$$F^\|(\mathbf{R} - \mathbf{R}') \;=\; \int_0^\beta d\tau \left\langle T_\tau \left[\varphi_\|(\mathbf{R}, \tau)\varphi_\|(\mathbf{R}', \tau = 0)\right]\right\rangle - \beta|\Delta|^2 \tag{3.5}$$

$$F^\perp(\mathbf{R} - \mathbf{R}') \;=\; \int_0^\beta d\tau \left\langle T_\tau \left[\varphi_\perp(\mathbf{R}, \tau)\varphi_\perp(\mathbf{R}', \tau = 0)\right]\right\rangle \tag{3.6}$$

where T_τ is the imaginary-time T-ordering operator, the thermal average $\langle\ldots\rangle$ is taken at the equilibrium temperature β^{-1}, and the Heisenberg representation for the field operators is implemented below. Note that the unnecessary dependence on the imaginary time τ has been eliminated in the expressions (3.5) by the time averaging.

For a homogeneous system, it is further convenient to introduce the Fourier transform of the operator (3.1):

$$\varphi(\mathbf{q}) = \frac{1}{\sqrt{\Omega}} \int d\mathbf{R}\; e^{-i\mathbf{q}\cdot\mathbf{R}} \varphi(\mathbf{R}) = \sum_{\mathbf{k}} \phi(\mathbf{k})c_\uparrow(\mathbf{k} + \mathbf{q}/2)c_\downarrow(-\mathbf{k} + \mathbf{q}/2) \tag{3.7}$$

where $\phi(\mathbf{k})$ is the Fourier transform of the function $\phi(\rho)$ of Eq. (3.1), $c_\sigma(\mathbf{k})$ is the destruction operator of wave vector $\mathbf{k}$ and spin σ, and Ω is the volume occupied by the system. In terms of the operator (3.7) we rewrite:

$$F^{\|,\perp}(\mathbf{R} - \mathbf{R}') = \pm \frac{1}{4\Omega} \sum_{\mathbf{q}} e^{i\mathbf{q}\cdot(\mathbf{R}-\mathbf{R}')} \int_0^\beta d\tau$$
$$\times \left\{\left\langle T_\tau[\varphi(\mathbf{q}, \tau)\varphi(-\mathbf{q}, 0) \pm \varphi(\mathbf{q}, \tau)\varphi(\mathbf{q}, 0)^\dagger]\right\rangle\right.$$
$$\left.\pm \left\langle T_\tau[\varphi(-\mathbf{q}, \tau)^\dagger\varphi(-\mathbf{q}, 0) + \varphi(-\mathbf{q}, \tau)^\dagger\varphi(\mathbf{q}, 0)^\dagger]\right\rangle\right\} - \frac{(1 \pm 1)}{2}\beta|\Delta|^2 \tag{3.8}$$

where the upper (lower) sign refers to $F^\|$ ($F^\perp$). Note that the τ-averaging selects the zero (Matsubara) frequency component of the correlation functions within braces in Eq. (3.8). Note also that in Eq. (3.8) we have eventually considered Δ to be real.

Below a critical temperature, one expects to identify a *finite* coherence length for longitudinal correlations only. In particular, the behavior for *small* $\mathbf{q}$ of the integrand in Eq. (3.8) is of interest whenever the correlation function $F^\|(\mathbf{R} - \mathbf{R}')$ has a well-behaved (exponential) spatial decay. Since the broken-symmetry condition resides in the *phase* of the order parameter (3.4), in the following we shall identify as ξ_{phase} the coherence length associated with $F^\|$.

Physically, ξ_{phase} provides an estimate of the spatial dimension over which the phase fluctuations are correlated. In the strong-coupling (BE) limit one thus expects ξ_{phase} to be much larger than the typical size of the fermionic pairs (which, in this limit, constitute truly bosonic entities). In this context, it is relevant to introduce an additional length (say, ξ_{pair}) which reduces to the size of the bound fermionic *pair* in the BE limit. On general ground, information on ξ_{pair} can be extracted from the fermionic pair-correlation function (with opposite spins)

$$g(\mathbf{r}) = \frac{1}{n^2} \left\langle \psi_\uparrow^\dagger(\mathbf{r})\psi_\downarrow^\dagger(0)\psi_\downarrow(0)\psi_\uparrow(\mathbf{r}) \right\rangle - \frac{1}{4} \qquad (3.9)$$

where n is the particle density and the constant Hartree term has been subtracted for convenience. For instance, at the mean-field level Eq. (3.9) becomes:

$$g(\mathbf{r}) = \frac{1}{n^2} \left| \left\langle \Phi | \psi_\uparrow^\dagger(\mathbf{r})\psi_\downarrow^\dagger(0) | \Phi \right\rangle \right|^2 \qquad (3.10)$$

where $|\Phi\rangle$ is the BCS ground state, and ξ_{pair} can be obtained as [39]

$$\xi_{pair}^2 = \frac{\int d\mathbf{r}\ g(\mathbf{r})\mathbf{r}^2}{\int d\mathbf{r}\ g(\mathbf{r})} . \qquad (3.11)$$

In the BE limit, ξ_{pair} obtained from Eq. (3.11) coincides with the bound-state radius of the associated two-fermion problem: At the mean-field level a *single* length enters the function (3.10) since no correlation is established between bound pairs. Beyond mean field, however, correlation between bound pairs should occur and the length ξ_{phase} should affect $g(\mathbf{r})$. Nonetheless, in the BE limit we expect the magnitudes of the two lengths ξ_{pair} and ξ_{phase} to be widely separated, in such a way that ξ_{pair} can still be extracted from $g(\mathbf{r})$ by inspection. For this reason, in the following we shall restrict in practice to the mean-field definition (3.11) with $g(\mathbf{r})$ given by Eq. (3.10). In the weak-coupling limit, on the other hand, we expect no difference between ξ_{phase} and ξ_{pair} (apart, possibly, from a trivial normalization factor due to the respective definitions). In other words, in the weak-coupling limit a *single* length characterizes the correlation within a Cooper pair *and* among different Cooper pairs (the correlation originating essentially from Pauli exclusion principle).

To proceed in the calculation of ξ_{phase} (and ξ_{pair}) we need a specific Hamiltonian to describe the interacting fermionic system. To connect with previous work on the crossover from BCS to BE, we adopt the model Hamiltonian considered by NSR:

$$H = \sum_{\mathbf{k},\sigma} \xi_\mathbf{k} c_{\sigma,\mathbf{k}}^{\dagger,\mathbf{k}} c_{\sigma,\mathbf{k}} + \sum_{\mathbf{k},\mathbf{k}',\mathbf{q}} V_{\mathbf{k},\mathbf{k}'} c_{\uparrow,\mathbf{k}+\mathbf{q}/2}^\dagger c_{\downarrow,-\mathbf{k}+\mathbf{q}/2}^\dagger c_{\downarrow,-\mathbf{k}'+\mathbf{q}/2} c_{\uparrow,\mathbf{k}'+\mathbf{q}/2} \qquad (3.12)$$

with $\xi_\mathbf{k} = \mathbf{k}^2/2m - \mu$ (μ being the chemical potential) [56]. This Hamiltonian differs from the usual BCS reduced Hamiltonian [30], in that it allows for finite values of the (center-of-mass) momentum $\mathbf{q}$ of the pair operator $c_\uparrow^\dagger c_\downarrow^\dagger$ while keeping the singlet spin pairing. Taking into account finite values of $\mathbf{q}$ is, in fact, necessary to represent the strong-coupling limit in terms of interacting bosons [57].

For convenience, we also take the (effective) attractive interaction potential in Eq. (3.12) of the separable form ($V < 0$):

$$V(\mathbf{k}, \mathbf{k}') = V w(\mathbf{k}) w(\mathbf{k}'). \tag{3.13}$$

In the usual BCS theory, $w(\mathbf{k}) = \theta(\epsilon_c - |\xi_\mathbf{k}|)$ specifies an abrupt cutoff about the Fermi surface (ϵ_c being the cutoff energy). To treat the strong-coupling limit on the same footing of the weak-coupling limit, $w(\mathbf{k})$ should instead interpolate smoothly between small and large $\mathbf{k}$. We take accordingly:

$$w(\mathbf{k}) = \left[1 + (\mathbf{k}/k_0)^2\right]^{-\gamma} \tag{3.14}$$

with $\gamma > 0$. We have verified that the restriction $1/4 < \gamma < 3/4$ ensures the relevant correlation functions to be well-defined via their Fourier transforms, as well as the bound-state radius for the associated two-body problem to vanish in the limit $|V| \to \infty$. In most of the following calculations we shall take the value $\gamma = 1/2$ considered by NSR [58].

3.1   Functional-Integral Approach

As discussed in the Introduction, although we are originally considering a system of interacting fermions, we are interested in treating properly the strong-coupling regime where the fermionic system gets mapped onto a system of interacting bosons. To this end, it is relevant to introduce for *any* coupling bosonic-like (collective) variables from the outset, which turn eventually into truly bosonic fields in the strong-coupling limit.

Functional integrals are especially suited for introducing collective variables and, at the same time, for providing one with conserving approximations even in the presence of condensates [35]. In this context, one obtains the relevant fermionic correlation functions by differentiating the generating functional [59]

$$Z[\bar{\eta}, \eta] = \frac{\int \mathcal{D}\bar{c}\mathcal{D}c \exp\{-S - S_{int}\}}{\int \mathcal{D}\bar{c}\mathcal{D}c \exp\{-S\}} \tag{3.15}$$

with respect to the "sources" η ($\bar{\eta}$), where

$$S = \int_0^\beta d\tau \left(\sum_{\mathbf{k},\sigma} \bar{c}_\sigma(\mathbf{k}, \tau) \frac{\partial}{\partial \tau} c_\sigma(\mathbf{k}, \tau) + H(\tau) \right) \tag{3.16}$$

and

$$S_{int} = \int_0^\beta d\tau \sum_{\mathbf{k},\sigma} \left(\bar{\eta}_\sigma(\mathbf{k}, \tau) c_\sigma(\mathbf{k}, \tau) + \bar{c}_\sigma(\mathbf{k}, \tau) \eta_\sigma(\mathbf{k}, \tau) \right). \tag{3.17}$$

In these expressions, $(c, \bar{c})$ and $(\eta, \bar{\eta})$ are Grassmann variables, and

$$H(\tau) = \sum_{\mathbf{k},\sigma} \xi_{\mathbf{k}} \bar{c}_\sigma(\mathbf{k}, \tau) c_\sigma(\mathbf{k}, \tau) + V \sum_{\mathbf{q}} \bar{\mathcal{B}}(\mathbf{q}, \tau) \mathcal{B}(\mathbf{q}, \tau) \qquad (3.18)$$

with

$$\mathcal{B}(\mathbf{q}, \tau) = \sum_{\mathbf{k}} w(\mathbf{k}) c_\downarrow(-\mathbf{k} + \mathbf{q}/2, \tau) c_\uparrow(\mathbf{k} + \mathbf{q}/2, \tau) \qquad (3.19)$$

is the Hamiltonian associated in the action with the operator (3.12) and the choice (3.13) for the interaction potential [60].

Bosonic-like variables are introduced via the Hubbard-Stratonovich transformation

$$\exp\left\{ -V \bar{\mathcal{B}}(\mathbf{q}, \tau) \mathcal{B}(\mathbf{q}, \tau) \right\} = -\frac{1}{\pi V} \int db^*(\mathbf{q}, \tau) db(\mathbf{q}, \tau)$$
$$\times \exp\left\{ \frac{1}{V} |b(\mathbf{q}, \tau)|^2 + b(\mathbf{q}, \tau) \bar{\mathcal{B}}(\mathbf{q}, \tau) + b^*(\mathbf{q}, \tau) \mathcal{B}(\mathbf{q}, \tau) \right\} \quad (3.20)$$

that holds for any $\mathbf{q}$ and τ, where the variables $b(\mathbf{q}, \tau)$ obey periodic boundary conditions $b(\mathbf{q}, \tau + \beta) = b(\mathbf{q}, \tau)$ [61]. It is further convenient to introduce time Fourier transforms for Grassmann and bosonic variables, and make the change of variables

$$\begin{aligned} \chi_1(k) &= \bar{c}_\uparrow(k) &,& \chi_2(k) &= c_\downarrow(-k) \\ \xi_1(k) &= -\bar{\eta}_\uparrow(k) &,& \xi_2(k) &= \eta_\downarrow(-k) \end{aligned} \qquad (3.21)$$

analogous to Nambu spinor transformation, with the short-hand notation $k = (\mathbf{k}, \omega_s)$ where $\omega_s = 2\pi(s + 1/2)\beta^{-1}$ (s integer) is a fermionic Matsubara frequency. The generating functional (3.15) can thus be rewritten in the form:

$$Z[\bar{\xi}, \xi] = \frac{\int \mathcal{D}\bar{\chi} \mathcal{D}\chi \mathcal{D}b^* \mathcal{D}b \exp\{-S' - S'_{int}\}}{\int \mathcal{D}\bar{\chi} \mathcal{D}\chi \mathcal{D}b^* \mathcal{D}b \exp\{-S'\}} \qquad (3.22)$$

where now

$$S' = -\frac{1}{\beta V} \sum_q |b(q)|^2 + \sum_{k,k'} (\bar{\chi}_1(k), \bar{\chi}_2(k)) \, \mathbf{M}(k, k') \begin{pmatrix} \chi_1(k') \\ \chi_2(k') \end{pmatrix} \qquad (3.23)$$

and

$$S'_{int} = \sum_k \sum_{i=1}^{2} \left(\bar{\xi}_i(k) \chi_i(k) + \bar{\chi}_i(k) \xi_i(k) \right). \qquad (3.24)$$

In Eq. (3.23), $q = (\mathbf{q}, \omega_\nu)$ where $\omega_\nu = 2\pi\nu\beta^{-1}$ (ν integer) is a bosonic Matsubara frequency, and $\mathbf{M}(k, k')$ is the 2×2 matrix

$$\mathbf{M}(k, k') = \begin{pmatrix} \epsilon(k)\delta_{k,k'} & , & \frac{1}{\beta}w(\frac{k+k'}{2})b^*(k - k') \\ \frac{1}{\beta}w(\frac{k+k'}{2})b(k' - k) & , & -\epsilon(-k)\delta_{k,k'} \end{pmatrix} \qquad (3.25)$$

with

$$\epsilon(k) = i\omega_s - \xi_{\mathbf{k}}. \qquad (3.26)$$

The Grassmann variables can be integrated out at this point in Eq. (3.22), yielding

$$Z[\bar{\xi}, \xi] = \frac{\int \mathcal{D}b^* \mathcal{D}b \exp\{-S_{eff} - S''_{int}\}}{\int \mathcal{D}b^* \mathcal{D}b \exp\{-S_{eff}\}} \tag{3.27}$$

where

$$S_{eff} = -\frac{1}{\beta V} \sum_q |b(q)|^2 - \text{tr} \ln \mathbf{M} \tag{3.28}$$

is the *effective* bosonic action and

$$S''_{int} = \sum_{k,k'} \sum_{i,i'=1}^{2} \bar{\xi}_i(k) \, \mathbf{M}^{-1}(k,k')_{i,i'} \, \xi_{i'}(k') \tag{3.29}$$

Note that the trace in Eq. (3.28) is performed over the four-momentum (k) and Nambu spin (i) indices, and that S_{eff} formally contains all powers in the bosonic variables b. .

To proceed further, one usually considers a quadratic (Gaussian) expansion of the effective action (3.28) in terms of $(b - b_0)$ where b_0 is a mean-field value. In particular, in Ref. 13 this procedure has been applied to derive the analog of the time-dependent Ginzburg-Landau equation for the crossover problem from BCS to BE above the mean-field critical temperature (where b_0 vanishes identically). For the zero-temperature properties we are interested in, however, a straightforward Gaussian expansion is known to be not fully consistent since it omits contributions that are formally of the same order of the Gaussian contribution itself (cf., e.g., Ref. 88 for the zero-temperature properties of a three-dimensional dilute Bose gas and Ref. 37 for its two-dimensional counterpart). To keep full consistency at each stage of the calculation, we introduce a (formal) *loop expansion* in the generating functional (3.27) by: (*i*) replacing the effective action S_{eff} with S_{eff}/λ where $0 < \lambda \leq 1$; (*ii*) regarding λ as the expansion parameter of the theory (to express, e.g., the correlation functions as power series in λ); (*iii*) setting $\lambda = 1$ eventually at the end of the calculation. In this way, expansion of the relevant physical quantities up to a given order in λ guarantees conservation laws and Ward identities to be satisfied to the same order in the expansion [62]. Note that, contrary to other cases for which a "small" loop parameter naturally emerges from the physics of the problem, the introduction of a loop parameter in the present context might at first look somewhat artificial. As mentioned in the Introduction, however, it can be shown that the present loop expansion gets formally mapped onto a *low-density* expansion in the bosonic limit [38].

To implement the loop expansion, we set

$$b(q) = \beta \left(\Delta_0 \delta_{q,0} + \sqrt{\lambda} \, \tilde{b}(q) \right) \tag{3.30}$$

where Δ_0 plays the role of a (complex) bosonic condensate and $\tilde{b}$ of its fluctuating part. The matrix (3.25) becomes accordingly:

$$\mathbf{M}_\lambda(k, k') = \mathbf{M}_0(k, k') + \sqrt{\lambda}\,\mathbf{M}_1(k, k') \tag{3.31}$$

where

$$\mathbf{M}_0(k, k') = \begin{pmatrix} \epsilon(k) & w(k)\Delta_0^* \\ w(k)\Delta_0 & -\epsilon(-k) \end{pmatrix} \delta_{k,k'} \tag{3.32}$$

and

$$\mathbf{M}_1(k, k') = w\left(\frac{k + k'}{2}\right)\begin{pmatrix} 0 & \tilde{b}^*(k - k') \\ \tilde{b}(k' - k) & 0 \end{pmatrix} \tag{3.33}$$

are independent of λ. Correspondingly, the effective action reads:

$$\frac{S_{eff}}{\lambda} = -\frac{\beta}{V}\left(\frac{|\Delta_0|^2}{\lambda} + \frac{\Delta_0}{\sqrt{\lambda}}\,\tilde{b}^*(q = 0) + \frac{\Delta_0^*}{\sqrt{\lambda}}\,\tilde{b}(q = 0) + \sum_q |\tilde{b}(q)|^2\right)$$
$$-\frac{1}{\lambda}\left(\mathrm{tr}\ln\mathbf{M}_0 - \sum_{n=1}^{\infty}\frac{(-1)^n}{n}\lambda^{n/2}\mathrm{tr}\left(\mathbf{M}_0^{-1}\mathbf{M}_1\right)^n\right). \tag{3.34}$$

The constant Δ_0 is determined, as usual, by requiring the coefficients of the linear terms in $\tilde{b}(q = 0)$ and $\tilde{b}^*(q = 0)$ to vanish, yielding the BCS "gap equation"

$$\Delta_0 = -V\sum_{\mathbf{k}}\frac{\Delta(\mathbf{k})w(\mathbf{k})}{2E(\mathbf{k})}\tanh\left(\frac{\beta E(\mathbf{k})}{2}\right) \tag{3.35}$$

with $\Delta(\mathbf{k}) = \Delta_0 w(\mathbf{k})$ and $E(\mathbf{k}) = \sqrt{\xi_{\mathbf{k}}^2 + |\Delta(\mathbf{k})|^2}$.

Equation (3.34) is still exact. Approximations depend on the number of power $\lambda^{n/2}$ considered. In particular, the Gaussian approximation for S_{eff} results upon keeping the next significant ($n = 2$) order in λ, namely, by taking

$$\frac{S_{eff}}{\lambda} = \frac{1}{\lambda}\beta F_0 + S_{eff}^{(2)} \tag{3.36}$$

where

$$F_0 = -\frac{|\Delta_0|^2}{V} - \frac{1}{\beta}\mathrm{tr}\ln\mathbf{M}_0 \tag{3.37}$$

is the (grand-canonical) free energy at the mean-field level [63] and $S_{eff}^{(2)}$ is the quadratic form

$$S_{eff}^{(2)} = \frac{1}{2}\sum_q(\tilde{b}^*(q), \tilde{b}(-q))\begin{pmatrix} A(q) & B(q) \\ B^*(q) & A(-q) \end{pmatrix}\begin{pmatrix} \tilde{b}(q) \\ \tilde{b}^*(-q) \end{pmatrix}. \tag{3.38}$$

In this expression:

$$A(q) = -\frac{\beta}{V} - \sum_k w(k - q/2)^2 \mathcal{G}(k)\mathcal{G}(q - k) = A(-q)^* \tag{3.39}$$

$$B(q) = \sum_k w(k - q/2)^2 \mathcal{F}(k)\mathcal{F}(q - k) = B(-q) \tag{3.40}$$

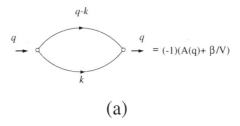

(a)

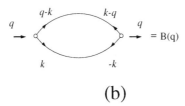

(b)

Figure 3.1: Normal (a) and anomalous (b) particle-particle bubbles. Full lines represent single-particle Green's functions and empty circles represent the function $w(k)$ of the separable potential (we have assumed $-w\left(\frac{k_1+k_2}{2}\right)$ to be even in its argument).

where

$$\mathcal{G}(k) = \mathbf{M}_0^{-1}(k)_{11} = \frac{\epsilon(-k)}{|\epsilon(k)|^2 + |\Delta(k)|^2} \tag{3.41}$$

$$\mathcal{F}(k) = \mathbf{M}_0^{-1}(k)_{21} = \frac{\Delta(k)}{|\epsilon(k)|^2 + |\Delta(k)|^2} \tag{3.42}$$

are the ordinary Gorkov functions. $A(q)+\beta/V$ and $B(q)$ represent normal and anomalous particle-particle bubbles, respectively, as depicted in Fig. 3.1 [64].

In what follows, it is sufficient to retain the quadratic action (3.36) only, *but* for the calculation of the shift Δ_1 of the mean-field parameter Δ_0 for which it is necessary to keep also cubic terms in the expansion (3.34) of the effective action (see Appendix A). In this way, the (grand-canonical) free energy acquires the following correction to the next significant order beyond mean field [65]:

$$F_1 = \frac{1}{2\beta} \sum_q \ln\left(|A(q)|^2 - B(q)^2\right) \tag{3.43}$$

where Δ_0 has been taken to be real and with the "stability" conditions

$$|A(q)| - B(q) > 0 , \qquad \Re A(q) + B(q) > 0 . \tag{3.44}$$

The chemical potential μ can be eventually eliminated in favor of the particle density n by solving $n = -(1/\Omega)\partial F/\partial \mu$ with $F = F_0 + \lambda F_1$. In principle, the chemical potential needs not be expanded in powers of λ since all conserving requirements can be directly expressed within the grand-canonical ensemble, the mapping between μ and n being established at the end of the calculation after having set $\lambda = 1$ in the expression for F. Nonetheless, one may alternatively regard μ as an internal parameter of the theory and expand it in series of λ at the outset (for details cf. Appendix C of Ref. 66). In the following, we calculate the physical quantities of interest keeping the value of μ unspecified, and expand μ in series of λ only in the final expressions.

3.2 Calculation of ξ_{phase} at the One-Loop Order

There remains to combine the calculation of the longitudinal ($F^{\parallel}$) and transverse ($F^{\perp}$) correlation functions (3.8) with the loop expansion.

For completeness, we report in the following the main steps of the calculation which might also serve for addressing additional correlation functions. For our specific purposes the relevant result is Eq. (3.2) below.

We need to relate first the broken-symmetry parameter (3.4) [cf. Eq. (3.7)]

$$\Delta = \frac{1}{\sqrt{\Omega}} \langle \varphi(\mathbf{q} = 0) \rangle = \frac{1}{\sqrt{\Omega}} \sum_{\mathbf{k}} \phi(\mathbf{k}) \langle c_{\uparrow}(\mathbf{k}) c_{\downarrow}(-\mathbf{k}) \rangle \tag{3.45}$$

with the mean-field value Δ_0 and the one-loop fluctuation contribution Δ_1. To this end, we rely on the identity (proven in Appendix A)

$$\langle b(q = 0) \rangle_{S_{eff}} = V \sum_{k} w(k) \langle c_{\uparrow}(k) c_{\downarrow}(-k) \rangle_{S} \tag{3.46}$$

where the averages are taken, respectively, with actions (3.28) and (3.16). Comparison of Eq. (3.46) with the definition (3.45) then yields

$$\Delta = \frac{1}{\beta} \langle b(q = 0) \rangle_{S_{eff}} = \Delta_0 + \sqrt{\lambda} \langle \tilde{b}(q = 0) \rangle_{S_{eff}} \tag{3.47}$$

with the notation (3.30) and the choice $\phi(\mathbf{k}) = \sqrt{\Omega} V w(\mathbf{k})$. The relation (3.47) is still exact. At the one-loop order it reduces to $\Delta = \Delta_0 + \lambda\Delta_1$, as shown in Appendix A.

Next, we express the averages of four-fermion operators in Eq. (3.8) (which are taken with action (3.16) within the functional-integral formulation) in terms of products of matrix elements of the inverse of the matrix (3.25) (which are correspondingly averaged with action (3.28)). We obtain:

$$F^{\parallel,\perp}(\mathbf{R}) = F_D^{\parallel,\perp}(\mathbf{R}) + F_E^{\parallel,\perp}(\mathbf{R}) - \frac{(1 \pm 1)}{2} \beta |\Delta|^2, \tag{3.48}$$

where

$$F_D^{\|,\perp}(\mathbf{R}) = \pm \frac{V^2}{4} \sum_{\mathbf{q}} e^{i\mathbf{q}\cdot\mathbf{R}} \sum_{\mathbf{k},\mathbf{k}'} w(\mathbf{k})w(\mathbf{k}') \frac{1}{\beta} \sum_{s,s'}$$

$$\times \Big\{ \Big\langle \mathbf{M}_{21}^{-1}(\mathbf{k}-\mathbf{q}/2,s;\mathbf{k}+\mathbf{q}/2,s)\mathbf{M}_{21}^{-1}(\mathbf{k}'+\mathbf{q}/2,s';\mathbf{k}'-\mathbf{q}/2,s') \Big\rangle_{S_{eff}}$$

$$\pm \Big\langle \mathbf{M}_{21}^{-1}(\mathbf{k}-\mathbf{q}/2,s;\mathbf{k}+\mathbf{q}/2,s)\mathbf{M}_{12}^{-1}(\mathbf{k}'+\mathbf{q}/2,s';\mathbf{k}'-\mathbf{q}/2,s') \Big\rangle_{S_{eff}}$$

$$\pm \Big\langle \mathbf{M}_{12}^{-1}(\mathbf{k}-\mathbf{q}/2,s;\mathbf{k}+\mathbf{q}/2,s)\mathbf{M}_{21}^{-1}(\mathbf{k}'+\mathbf{q}/2,s';\mathbf{k}'-\mathbf{q}/2,s') \Big\rangle_{S_{eff}}$$

$$+ \Big\langle \mathbf{M}_{12}^{-1}(\mathbf{k}-\mathbf{q}/2,s;\mathbf{k}+\mathbf{q}/2,s)\mathbf{M}_{12}^{-1}(\mathbf{k}'+\mathbf{q}/2,s';\mathbf{k}'-\mathbf{q}/2,s') \Big\rangle_{S_{eff}} \Big\}$$

$$(3.49)$$

is the "direct" contribution, and

$$F_E^{\|,\perp}(\mathbf{R}) = \mp \frac{V^2}{4} \sum_{\mathbf{q}} e^{i\mathbf{q}\cdot\mathbf{R}} \sum_{\mathbf{k},\mathbf{k}'} w(\mathbf{k})w(\mathbf{k}') \frac{1}{\beta} \sum_{s,s'}$$

$$\times \Big\{ \Big\langle \mathbf{M}_{21}^{-1}(\mathbf{k}-\mathbf{q}/2,s;\mathbf{k}'-\mathbf{q}/2,s')\mathbf{M}_{21}^{-1}(\mathbf{k}'+\mathbf{q}/2,s';\mathbf{k}+\mathbf{q}/2,s) \Big\rangle_{S_{eff}}$$

$$\pm \Big\langle \mathbf{M}_{22}^{-1}(\mathbf{k}-\mathbf{q}/2,s;\mathbf{k}'-\mathbf{q}/2,s')\mathbf{M}_{11}^{-1}(\mathbf{k}'+\mathbf{q}/2,s';\mathbf{k}+\mathbf{q}/2,s) \Big\rangle_{S_{eff}}$$

$$\pm \Big\langle \mathbf{M}_{11}^{-1}(\mathbf{k}-\mathbf{q}/2,s;\mathbf{k}'-\mathbf{q}/2,s')\mathbf{M}_{22}^{-1}(\mathbf{k}'+\mathbf{q}/2,s';\mathbf{k}+\mathbf{q}/2,s) \Big\rangle_{S_{eff}}$$

$$+ \Big\langle \mathbf{M}_{12}^{-1}(\mathbf{k}-\mathbf{q}/2,s;\mathbf{k}'-\mathbf{q}/2,s')\mathbf{M}_{12}^{-1}(\mathbf{k}'+\mathbf{q}/2,s';\mathbf{k}+\mathbf{q}/2,s) \Big\rangle_{S_{eff}} \Big\}$$

$$(3.50)$$

is the "exchange" counterpart.

The loop expansion emerges at this point from the exact expressions (3.49) and (3.50) by interpreting the matrix $\mathbf{M}$ therein as being the matrix $\mathbf{M}_\lambda$ (3.31). In this way, its inverse acquires the expansion

$$\mathbf{M}_{\alpha\beta}^{-1} = (\mathbf{M}_0^{-1})_{\alpha\beta} - \sqrt{\lambda}\left(\mathbf{M}_0^{-1}\mathbf{M}_1\mathbf{M}_0^{-1}\right)_{\alpha\beta} + \lambda\left(\mathbf{M}_0^{-1}\mathbf{M}_1\mathbf{M}_0^{-1}\mathbf{M}_1\mathbf{M}_0^{-1}\right)_{\alpha\beta} + \cdots,$$

$$(3.51)$$

yielding for the required averages

$$\Big\langle \mathbf{M}_{\alpha\beta}^{-1}\mathbf{M}_{\gamma\delta}^{-1} \Big\rangle = \Big\langle \mathbf{M}_{\alpha\beta}^{-1} \Big\rangle \Big\langle \mathbf{M}_{\gamma\delta}^{-1} \Big\rangle$$

$$+ \lambda \Big\langle (\mathbf{M}_0^{-1}\mathbf{M}_1\mathbf{M}_0^{-1})_{\alpha\beta}(\mathbf{M}_0^{-1}\mathbf{M}_1\mathbf{M}_0^{-1})_{\gamma\delta} \Big\rangle + \mathcal{O}(\lambda^{3/2}) \qquad (3.52)$$

with the understanding that the product $\Big\langle \mathbf{M}_{\alpha\beta}^{-1} \Big\rangle \Big\langle \mathbf{M}_{\gamma\delta}^{-1} \Big\rangle$ is evaluated at the relevant order in λ. [In the expressions above, the indices α, β, ... refer to the four-vector k and the Nambu spinor component]. In particular, *at the mean-field level* Eq. (3.52) reduces to

$$\Big\langle \mathbf{M}_{\alpha\beta}^{-1}\mathbf{M}_{\gamma\delta}^{-1} \Big\rangle \to (\mathbf{M}_0^{-1})_{\alpha\beta}(\mathbf{M}_0^{-1})_{\gamma\delta} \qquad (3.53)$$

with $\mathbf{M}_0$ given by Eq. (3.32) (and Δ_0 taken eventually to be real). In this case the "direct" contribution (3.49) becomes:

$$F_D^{\|,\perp}(\mathbf{R}) \to \frac{(1 \pm 1)}{2} \beta \left(\frac{V}{\beta} \sum_k w(k) \mathbf{M}_0^{-1}(k)_{21} \right)^2 = \frac{(1 \pm 1)}{2} \beta \Delta_0^2 \qquad (3.54)$$

where use has been made of the gap equation (3.35). This contribution cancels the last term on the right-hand side of Eq. (3.48) since $\Delta \to \Delta_0$ at the mean-field level. On the other hand, the "exchange" contribution (3.50) becomes:

$$F_E^{\|,\perp}(\mathbf{R}) \to -\frac{V^2}{2\beta} \sum_{\mathbf{q}} e^{i\mathbf{q}\cdot\mathbf{R}} \left\{ A(\mathbf{q}, \omega_\nu = 0) \pm B(\mathbf{q}, \omega_\nu = 0) \right\} \qquad (3.55)$$

where $A(q)$ and $B(q)$ are given by Eqs. (3.39) and (3.40), respectively. [Equation (3.55) holds apart from a local term proportional to the delta function of argument $\mathbf{R}$, which is consistently neglected in the following]. In particular, in the zero-temperature limit Eq. (3.55) can be cast in the form:

$$F_E^{\|,\perp}(\mathbf{R}) \to \frac{V^2}{4} \sum_{\mathbf{q}} e^{i\mathbf{q}\cdot\mathbf{R}} \sum_{\mathbf{k}} \frac{w(\mathbf{k} - \mathbf{q}/2)^2}{E_{\mathbf{k}} + E_{\mathbf{k}-\mathbf{q}}} \left[1 + \frac{\xi_{\mathbf{k}}\xi_{\mathbf{k}-\mathbf{q}} \mp \Delta_{\mathbf{k}}\Delta_{\mathbf{k}-\mathbf{q}}}{E_{\mathbf{k}} E_{\mathbf{k}-\mathbf{q}}} \right] \qquad (3.56)$$

where the function to be summed over $\mathbf{q}$ is well-behaved for all $\mathbf{q}$ and contributes a "short-range" function of $\mathbf{R}$. Expression (3.56) will be studied numerically in Chapter 5 to determine its range explicitly.

The relevant "long-range" behavior in $\mathbf{R}$ results instead *at the one-loop level*. Entering Eq. (3.52) into Eq. (3.49), we obtain for the "direct" contribution:

$$\begin{aligned} F_D^{\|,\perp}(\mathbf{R}) = {} & \frac{(1 \pm 1)}{2}\beta\Delta^2 + \lambda\frac{V^2}{2\beta} \sum_{\mathbf{q}} e^{i\mathbf{q}\cdot\mathbf{R}} \left\{ A(\mathbf{q}, \omega_\nu = 0) \pm B(\mathbf{q}, \omega_\nu = 0) \right\} \\ & + \frac{\lambda}{2} \sum_{\mathbf{q}} e^{i\mathbf{q}\cdot\mathbf{R}} \frac{\beta}{A(\mathbf{q}, \omega_\nu = 0) \pm B(\mathbf{q}, \omega_\nu = 0)} \end{aligned} \qquad (3.57)$$

(apart, again, from a term proportional to a delta function of $\mathbf{R}$). Note that the first term on the right-hand side of Eq. (3.57) results from the first term on the right-hand side of Eq. (3.52) together with Eqs. (3.46) and (3.47) (cf. Appendix A). In this term Δ is meant to contain also its one-loop shift Δ_1 (which is real when Δ_0 is real), making it to cancel with the last term of Eq. (3.48). Note further that the second term on the right-hand side of Eq. (3.57) coincides formally (apart from a sign) with the mean-field contribution (3.55) once one sets $\lambda = 1$, and thus shares the same "short-range" character. The last term on the right-hand side of Eq. (3.57), on the other hand, yields the desired "long-range" behavior.

Before discussing this behavior in detail, it is worth to represent graphically expression (3.49) at the order of the approximation (3.52). This is done in

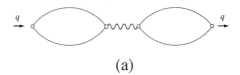

(a)

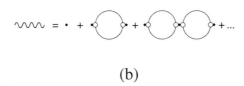

(b)

Figure 3.2: (*a*) Graphical representation of a typical "direct" term of order λ in Eq. (3.49) *before* it is integrated over the wave vector $\mathbf{q}$ (recall that, by our definition (3.5) of the correlation functions, $q = (\mathbf{q}, \omega_\nu = 0)$ in this term). For simplicity, arrows distinguishing normal and anomalous single-particle Green's functions are not indicated. The wiggly line stands for the (transposed) matrix of the bosonic propagator $\left\langle \tilde{\mathbf{b}}(q)\tilde{\mathbf{b}}^\dagger(q) \right\rangle_{S_{eff}^{(2)}}$, where $\tilde{\mathbf{b}}(q)$ is the column vector of Eq. (3.38). This propagator is depicted in (*b*) (at the order considered in the present work) as an infinite series of the original fermionic bubbles (dots represent the strength V of the separable potential.)

Fig. 3.2 for the terms of order λ. It is evident from the figure that the bosonic propagator (wiggly line) carries the external (four) momentum q, such that any singularity of this propagator for small values of q will affect the spatial decay of the "direct" contribution (3.49) to the correlation functions.

By contrast, in the "exchange" contribution (3.50) the bosonic propagator does not carry the external (four) momentum q since this propagator occurs entangled in the internal structure of the diagrams (cf. Fig. 3.3). In this case the singularity of the propagator for small momenta is smoothed out by the internal (four) momentum integrations in the diagrams. For this reason, the "exchange" diagrams are not expected to contribute to the "long-range" behavior of the correlation functions (3.48) and are accordingly neglected in the following [67].

In conclusion, at the order λ (one-loop) we approximate the correlation functions (3.5) by the following expressions:

$$F^\parallel(\mathbf{R}) \;\cong\; \frac{\lambda}{2} \sum_{\mathbf{q}} e^{i\mathbf{q}\cdot\mathbf{R}} \frac{\beta}{A(\mathbf{q}, \omega_\nu = 0) + B(\mathbf{q}, \omega_\nu = 0)} \qquad (3.58)$$

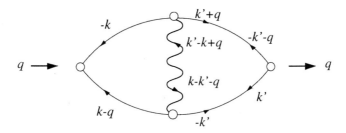

Figure 3.3: Typical diagram of order λ occurring in the "exchange" contribution (3.50). The internal (four) momenta k and k' are integrated, and the momentum $q = (\mathbf{q}, \omega_\nu = 0)$ is associated with the spatial ($\mathbf{R}$) dependence. Note that diagrams of this type vanish in the "normal" phase (when $\Delta_0 \to 0$).

$$F^\perp(\mathbf{R}) \cong \frac{\lambda}{2} \sum_{\mathbf{q}} e^{i\mathbf{q}\cdot\mathbf{R}} \frac{\beta}{A(\mathbf{q}, \omega_\nu = 0) - B(\mathbf{q}, \omega_\nu = 0)} \qquad (3.59)$$

with $A(q)$ and $B(q)$ given by Eqs. (3.39) and (3.40), respectively. Note that in the normal phase (for which $B(q) = 0$) there is no distinction between longitudinal and transverse correlation functions. In the superconducting phase, on the other hand, a coherence length can be identified only for the longitudinal correlation function, as expected. This is because

$$\lim_{\mathbf{q}\to 0}\left(A(\mathbf{q}, \omega_\nu = 0) - B(\mathbf{q}, \omega_\nu = 0)\right) = -\beta\left[\frac{1}{V} + \sum_{\mathbf{k}} \frac{w(\mathbf{k})^2}{2E_\mathbf{k}} \tanh\left(\frac{\beta E_\mathbf{k}}{2}\right)\right]$$
$$(3.60)$$

vanishes owing to the gap equation (3.35). The other combination $A(\mathbf{q}, \omega_\nu = 0) + B(\mathbf{q}, \omega_\nu = 0)$ is instead finite for $\mathbf{q} \to 0$.

If we restrict, in particular, to the *zero-temperature limit*, we obtain:

$$A(\mathbf{q}, \omega_\nu = 0) + B(\mathbf{q}, \omega_\nu = 0)$$
$$= -\beta\left[\frac{1}{V} + \sum_{\mathbf{k}} w(\mathbf{k} - \mathbf{q}/2)^2 \frac{(u_\mathbf{k}u_{\mathbf{k}-\mathbf{q}} - v_\mathbf{k}v_{\mathbf{k}-\mathbf{q}})^2}{E_\mathbf{k} + E_{\mathbf{k}-\mathbf{q}}}\right] \equiv \beta f(\mathbf{q}) \quad (3.61)$$

where

$$u_\mathbf{k} = \sqrt{\frac{1}{2}\left(1 + \frac{\xi_\mathbf{k}}{E_\mathbf{k}}\right)}, \qquad v_\mathbf{k} = \sqrt{\frac{1}{2}\left(1 - \frac{\xi_\mathbf{k}}{E_\mathbf{k}}\right)} \qquad (3.62)$$

are the usual BCS parameters. Note that $f(\mathbf{q}) = f(|\mathbf{q}|)$ and

$$\lim_{\mathbf{q}\to 0} f(\mathbf{q}) = a = \sum_{\mathbf{k}} w(\mathbf{k})^2 \frac{\Delta_\mathbf{k}^2}{2E_\mathbf{k}^3} > 0 \qquad (3.63)$$

provided $\Delta_0 \neq 0$, owing again to the gap equation (3.35). For small values of $\mathbf{q}$ we can thus expand

$$f(\mathbf{q}) = a + b\mathbf{q}^2 + \ldots \qquad , \qquad (3.64)$$

and obtain the desired coherence length as follows:

$$\xi_{phase} = \sqrt{\frac{b}{a}} \qquad (3.65)$$

provided b is also positive. In fact, entering the expansion (3.64) into Eq. (3.58) yields for the leading "long-range" behavior (in three dimensions):

$$F^{\parallel}(\mathbf{R}) \approx \frac{\lambda}{2} \sum_{\mathbf{q}} e^{i\mathbf{q}\cdot\mathbf{R}} \frac{1}{a + b\mathbf{q}^2} = \frac{\lambda}{2} \frac{\Omega}{4\pi b} \frac{\exp\{-|\mathbf{R}|/\xi_{phase}\}}{|\mathbf{R}|} \qquad (3.66)$$

with ξ_{phase} given by Eq. (3.65). Consideration of the expansion (3.64) is obviously sufficient provided the function $f(\mathbf{q})$ has no other singularity [68].

There remains to obtain an explicit expression for the coefficient b of the expansion (3.64), for generic values of the parameters characterizing the interaction potential (3.13). To this end, we expand Eq. (3.61) retaining all terms up to order $\mathbf{q}^2$ and obtain (in three dimensions):

$$b = \frac{1}{8m} \sum_{\mathbf{k}} \frac{w(\mathbf{k})^2 \xi_{\mathbf{k}}^2}{E_{\mathbf{k}}^5} \left\{ \frac{(\xi_{\mathbf{k}}^2 - 2\Delta_{\mathbf{k}}^2)}{2\xi_{\mathbf{k}}} + \frac{3}{2} z'(\mathbf{k})\Delta_{\mathbf{k}}^2 \right.$$
$$\left. + \frac{\mathbf{k}^2}{6m} \frac{\Delta_{\mathbf{k}}^2}{\xi_{\mathbf{k}} E_{\mathbf{k}}^2} \left[5\xi_{\mathbf{k}} + z'(\mathbf{k})(4\Delta_{\mathbf{k}}^2 - 6\xi_{\mathbf{k}}^2) + z'(\mathbf{k})^2 \xi_{\mathbf{k}}(\xi_{\mathbf{k}}^2 - 4\Delta_{\mathbf{k}}^2) + 3z''(\mathbf{k})\xi_{\mathbf{k}} E_{\mathbf{k}}^2 \right] \right\}$$
$$(3.67)$$

with the notation

$$z'(\mathbf{k}) = \frac{2m}{w(\mathbf{k})} \frac{dw(\mathbf{k})}{d\mathbf{k}^2} , \qquad z''(\mathbf{k}) = \frac{(2m)^2}{w(\mathbf{k})} \frac{d^2 w(\mathbf{k})}{d(\mathbf{k}^2)^2} . \qquad (3.68)$$

Quite generally, all terms within braces in Eq. (3.67) contribute to the value of b for intermediate coupling and the sum over the wave vector has correspondingly to be evaluated numerically. This task will be performed in Chapter 5. In the extreme (weak- and strong-coupling) limits, on the other hand, only a single term within braces in Eq. (3.67) (albeit different in the two cases) contributes to the value of b and the sum over the wave vector can be evaluated analytically.

3.3   Analytic Results in the BCS and BE Limits

It is worth to show in detail how the coefficients a and b of the expansion (3.64) can be evaluated *analytically* in the BCS and BE limits, by exploiting simplifying features of the calculation. Specifically, the sum over the wave vector in Eqs. (3.63) and (3.67) will be evaluated with the approximation $\Delta_0/|\mu| \ll 1$

that holds in both limits (albeit with $\mu > 0$ and $\mu < 0$, respectively).   The main results of this Chapter are given by Eqs. (3.71) and (3.81) below.

In the weak-coupling (BCS) limit the term $(5/3)(\mathbf{k}^2/2m)(\Delta_{\mathbf{k}}/E_{\mathbf{k}})^2$ within braces in Eq. (3.67) provides the dominant contribution [69], yielding (in three dimensions):

$$
\begin{aligned}
b_{BCS} &= \frac{5}{24m} \sum_{\mathbf{k}} w(\mathbf{k})^2 \frac{\mathbf{k}^2}{2m} \frac{\xi_{\mathbf{k}}^2 \Delta_{\mathbf{k}}^2}{E_{\mathbf{k}}^7} \\
&= \Omega \frac{5k_F w(k_F)^2}{48\pi^2 \mu \tilde{\Delta}_0^2} \int_{-\tilde{\Delta}_0^{-1}}^{+\infty} dy \frac{\tilde{w}(y)^4 (y\tilde{\Delta}_0 + 1)^{3/2} y^2}{[y^2 + \tilde{w}(y)^2]^{7/2}} \\
&\cong \Omega \frac{5k_F w(k_F)^2}{48\pi^2 \mu \tilde{\Delta}_0^2} \int_{-\infty}^{+\infty} dy \frac{y^2}{(y^2 + 1)^{7/2}} \\
&= \Omega \frac{m k_F w(k_F)^2}{2\pi^2} \frac{1}{(3k_F \tilde{\Delta}_0)^2} .
\end{aligned} \tag{3.69}
$$

In the expression above, we have set $w(k) = w(k_F)\tilde{w}(\tilde{k})$ with $\tilde{k} = k/k_F$ and $\tilde{w}(\tilde{k} = 1) = 1$, $\xi_k/\mu = \tilde{\xi}_{\tilde{k}} = \tilde{k}^2 - 1$ (since $\mu = k_F^2/2m$ in the BCS limit), $\Delta_k/\mu = \tilde{w}(\tilde{k})\tilde{\Delta}_0$ with $\tilde{\Delta}_0 = w(k_F)\Delta_0/\mu = \Delta_{k_F}/\mu$, and $y = \tilde{\xi}/\tilde{\Delta}_0$. In addition, the last line of Eq. (3.69) has been obtained by exploiting the BCS condition $\tilde{\Delta}_0 \ll 1$ as well as the normalization $\tilde{w}(y = 0) = \tilde{w}(\tilde{k} = 1) = 1$. By the same token, the coefficient a given by Eq. (3.63) becomes in the BCS limit (in three dimensions):

$$
\begin{aligned}
a_{BCS} &= \sum_{\mathbf{k}} w(\mathbf{k})^2 \frac{\Delta_{\mathbf{k}}^2}{2E_{\mathbf{k}}^3} \\
&= \Omega \frac{m k_F w(k_F)^2}{4\pi^2} \int_{-\tilde{\Delta}_0^{-1}}^{+\infty} dy \frac{\tilde{w}(y)^4 (y\tilde{\Delta}_0 + 1)^{1/2}}{[y^2 + \tilde{w}(y)^2]^{3/2}} \\
&\cong \Omega \frac{m k_F w(k_F)^2}{4\pi^2} \int_{-\infty}^{+\infty} dy \frac{1}{(y^2 + 1)^{3/2}} \\
&= \Omega \frac{m k_F w(k_F)^2}{2\pi^2} .
\end{aligned} \tag{3.70}
$$

Entering the results (3.69) and (3.70) into Eq. (3.65), we obtain eventually:

$$
\xi_{phase}^{BCS} = \sqrt{\frac{b_{BCS}}{a_{BCS}}} = \frac{1}{3k_F \tilde{\Delta}_0} . \tag{3.71}
$$

This result has to be compared with the BCS limit for ξ_{pair} (cf. Eqs. (3.10) and (3.11)) obtained previously [39], namely,

$$
\xi_{pair}^{BCS} = \frac{1}{\sqrt{2}k_F \tilde{\Delta}_0} . \tag{3.72}
$$

Apart from a numerical factor of order unity due to a different normalization in the respective definitions, ξ_{phase} is thus seen to coincide with ξ_{pair} in the (extreme) BCS limit, as expected. What is less obviously expected, however, is the fact that the ratio ξ_{phase}/ξ_{pair} maintains its BCS value $\sqrt{2}/3 \simeq 0.47$ not only asymptotically (i.e., for $k_F\xi_{pair} \approx 10^3 - 10^4$) but *also* down to $k_F\xi_{pair} \simeq 10$ where bosonization starts to occur, as we shall verify in Chapter 5 by calculating the expressions (3.11) and (3.65) numerically.

In the opposite strong-coupling (BE) limit, the term $\xi_{\mathbf{k}}/2$ within braces in Eq. (3.67) provides instead the dominant contribution to the coefficient b [70], yielding

$$b_{BE} = \frac{1}{8m}\sum_{\mathbf{k}} \frac{w(\mathbf{k})^2}{2\xi_{\mathbf{k}}^{0\,2}} \tag{3.73}$$

with $\xi_{\mathbf{k}}^0 = \mathbf{k}^2/2m + \epsilon_0/2$ [$-\epsilon_0$ being the (lowest) eigenvalue of the associated eigenvalue problem for two fermions interacting via the potential (3.13)]. By the same token, we obtain for the coefficient a in the BE limit (cf. Eq. (3.63)):

$$a_{BE} = \sum_{\mathbf{k}} w(\mathbf{k})^2 \frac{\Delta_{\mathbf{k}}^2}{2\xi_{\mathbf{k}}^{0\,3}}. \tag{3.74}$$

Upon taking the ratio

$$(\xi_{phase}^{BE})^2 = \frac{b_{BE}}{a_{BE}} = \frac{1}{8m}\frac{\sum_{\mathbf{k}}\frac{w(\mathbf{k})^2}{\xi_{\mathbf{k}}^{0\,2}}}{\sum_{\mathbf{k}}w(\mathbf{k})^2\frac{\Delta_{\mathbf{k}}^2}{\xi_{\mathbf{k}}^{0\,3}}} \equiv \frac{1}{8\mu_1(2m)}, \tag{3.75}$$

we recognize the quantity μ_1 to be the (positive) shift of the chemical potential (at the lowest significant order in Δ_0/ϵ_0) with respect to the asymptotic value $\mu_0 = -\epsilon_0/2 < 0$. To show this, we resort to the mean-field equation (3.35) in the zero-temperature limit

$$1 + V\sum_{\mathbf{k}} \frac{w(\mathbf{k})^2}{2E_{\mathbf{k}}} = 0 \tag{3.76}$$

and expand in Δ_0/ϵ_0

$$\begin{aligned}E_{\mathbf{k}} &= \xi_{\mathbf{k}}\left[1 + \frac{\Delta_{\mathbf{k}}^2}{2\xi_{\mathbf{k}}^2} + \cdots\right]\\ &= \left(\frac{\mathbf{k}^2}{2m} - \mu_0 - \mu_1 + \cdots\right)\left[1 + \frac{1}{2}\frac{\Delta_{\mathbf{k}}^2}{\left(\frac{\mathbf{k}^2}{2m} - \mu_0 - \mu_1 + \cdots\right)^2} + \cdots\right]\\ &\simeq \left(\frac{\mathbf{k}^2}{2m} - \mu_0\right) + \frac{1}{2}\frac{\Delta_{\mathbf{k}}^2}{\left(\frac{\mathbf{k}^2}{2m} - \mu_0\right)} - \mu_1,\end{aligned} \tag{3.77}$$

where μ_0 is solution to the bound-state equation

$$1 + V \sum_k \frac{w(\mathbf{k})^2}{\frac{k^2}{m} - 2\mu_0} = 0 \tag{3.78}$$

that gives $\mu_0 = -\epsilon_0/2$. Entering the approximation (3.77) into Eq. (3.76) and expanding further $E_\mathbf{k}^{-1}$ at the relevant order yield:

$$\begin{aligned}
0 &= 1 + V \sum_k \frac{w(\mathbf{k})^2}{2\xi_\mathbf{k}^0} \left[1 - \frac{1}{2}\frac{\Delta_\mathbf{k}^2}{\xi_\mathbf{k}^{0\,2}} + \frac{\mu_1}{\xi_\mathbf{k}^0} + \cdots\right] \\
&= -\frac{V}{4} \sum_k w(\mathbf{k})^2 \frac{\Delta_\mathbf{k}^2}{\xi_\mathbf{k}^{0\,3}} + \mu_1 \frac{V}{2} \sum_k \frac{w(\mathbf{k})^2}{\xi_\mathbf{k}^{0\,2}} + \cdots
\end{aligned} \tag{3.79}$$

where use has been made of Eq. (3.78). Solving for the shift μ_1 we obtain eventually:

$$\mu_1 = \frac{1}{2} \frac{\sum_k w(\mathbf{k})^2 \frac{\Delta_\mathbf{k}^2}{\xi_\mathbf{k}^{0\,3}}}{\sum_k \frac{w(\mathbf{k})^2}{\xi_\mathbf{k}^{0\,2}}} \tag{3.80}$$

as anticipated in Eq. (3.75).

The expression (3.75) coincides formally with the (square of the) coherence length associated with a truly bosonic system in the limit of weak boson-boson interaction (or low density), whereby $2\mu_1 = v(0)n_B$ ($v(0)$ being the zero-momentum component of the boson-boson interaction and n_B the bosonic density). In Chapter 4 we shall obtain an explicit expression for the *residual* boson-boson interaction which results upon bosonization of the original fermionic system, and verify that the product of its zero-momentum component times the bosonic density coincides with the expression (3.80) for $2\mu_1$ (at least, at the one-loop order we are considering in this Thesis). We regard this result as being a rather compelling check on our one-loop calculation to provide a consistent description of the dilute *interacting* Bose gas, obtained through bosonization of the original fermionic system.

The above results hold for any reasonable choice of the function $w(\mathbf{k})$. With the specific form (3.14) and $\gamma = 1/2$, the integrals occurring in Eqs. (3.78) and (3.80) can be performed analytically, yielding (in three dimensions)

$$(k_F \xi_{phase}^{BE})^2 = \frac{3\pi}{64} \tilde{k}_0 f(c) , \tag{3.81}$$

where $\tilde{k}_0 = k_0/k_F$, $c = [|\mu_0|/(k_0^2/2m)]^{1/2}$, and

$$f(c) = \frac{4c}{1 + 4c} \tag{3.82}$$

is a monotonically increasing function of c bounded between $f(c = 0) = 0$ and $f(c = \infty) = 1$. In particular, $f(c = 0.1) \simeq 0.286$. The appropriate value of

c (to be inserted in Eq. (3.81)) is then provided by Eq. (3.78), which in the present context reads

$$c = -\frac{V\Omega m k_0}{4\pi} - 1. \tag{3.83}$$

By our assumptions on how the BE limit is achieved (cf. Ref. 70), the value of c is expected to be much smaller than unity, and thus $f(c)$ to be at most of the order 1/3, yielding $\sqrt{\pi \tilde{k}_0}/8$ for the maximum attainable value of $k_F \xi_{phase}^{BE}$.

The dependence of ξ_{phase}^{BE} on the interaction strength should be contrasted with the value of ξ_{pair} obtained previously in the BE limit (namely $\xi_{pair}^{BE} = r_0$, r_0 being the mean radius of the bound-fermion pair) [39], such that $\xi_{phase}^{BE} \gg \xi_{pair}^{BE}$ for sensible choices of k_0 [71]. Again, this result is consistent with what we had expected in the bosonic limit, where the "internal" size r_0 of the bosons represents the smallest length in the problem and is certainly not related with the distance over which the fluctuations of the order parameter correlate.

Chapter 4

Mapping onto a Bosonic System in the Strong-Coupling Limit

One of the advantages for using the functional-integral approach in the crossover from BCS to BE is that it allows for a *direct* mapping of the original fermionic system in the strong-coupling limit onto an *interacting* bosonic system, at the level of the effective action. The Hubbard-Stratonovich decoupling (3.20) has, in fact, resulted into the effective bosonic action (3.28) for the boson-like complex variables $b(q)$, which resembles a truly bosonic action albeit with an infinite number of couplings. Some caution, however, is in order since the single-particle propagator associated with $b(q)$ lacks the characteristic equal-time step singularity, that is expected for a bosonic propagator owing to the bosonic commutator $[b, b^{\dagger}] = 1$. For this reason it will be necessary to reinterpret appropriately the field $b(q)$ in Eq. (3.28), in order to recover a truly bosonic action in the strong coupling limit.

Purpose of this Chapter is to carry out in detail the mapping onto a bosonic system *at the level of the effective action*, so as to obtain the residual (quartic) interaction among the composite bosons constituted by fermionic pairs [72]. This method will enable us to obtain (at the one-loop order) the phase coherence length for the limiting bosonic system directly in terms of the parameters of the residual interaction, and to compare it with the result obtained for the BE limit in Chapter 3. In this way, we will recover the expression for the coherence length of an interacting *dilute* Bose gas [33], thus establishing a consistency check on the approach of Chapter 3.

We begin by considering again the effective bosonic action (3.28) and expand $\operatorname{tr} \ln \mathbf{M}$ therein in powers of $b(q)$ about $b = 0$, rather than about the broken-symmetry value $b_0 = \beta \Delta_0$ as we did in Eq. (3.30). In addition, we shall not introduce here the loop parameter λ since it is not relevant to the following arguments. We thus split

$$\mathbf{M}(k, k') = \mathbf{M}_S(k, k') + \mathbf{M}_R(k, k') \tag{4.1}$$

41

where now

$$\mathbf{M}_S(k, k') = \begin{pmatrix} \epsilon(k) & 0 \\ 0 & -\epsilon(-k) \end{pmatrix} \delta_{k,k'} \tag{4.2}$$

and

$$\mathbf{M}_R(k, k') = w\left(\frac{k + k'}{2}\right)\begin{pmatrix} 0 & \frac{1}{\beta}b^*(k - k') \\ \frac{1}{\beta}b(k' - k) & 0 \end{pmatrix} \tag{4.3}$$

[cf. Eqs. (3.31)-(3.33)], and obtain

$$S_{eff} = -\text{tr}\ln\mathbf{M}_S - \frac{1}{\beta V}\sum_q |b(q)|^2 + \sum_{n=1}^{\infty}\frac{1}{2n}\text{tr}X^{2n} \tag{4.4}$$

with $X = \mathbf{M}_S^{-1}\mathbf{M}_R$:

$$X(k, k') = w\left(\frac{k + k'}{2}\right)\begin{pmatrix} 0 & \frac{b^*(k-k')}{\beta\epsilon(k)} \\ -\frac{b(k'-k)}{\beta\epsilon(-k)} & 0 \end{pmatrix}. \tag{4.5}$$

We retain first the quadratic terms in Eq. (4.4), that give

$$S_{eff}^{(2)} = \frac{1}{\beta^2}\sum_q |b(q)|^2\left[-\frac{\beta}{V} - \sum_k w(k - q/2)^2\frac{1}{\epsilon(k)\epsilon(q-k)}\right] \tag{4.6}$$

in the place of Eq. (3.38), the expression within brackets coinciding with $A(q)$ given by Eq. (3.39) in the limit $\Delta_0 = 0$. Keeping the same notation and performing the sum over the fermionic Matsubara frequencies, we obtain:

$$A(\mathbf{q}, z) = -\frac{\beta}{V} - \frac{\beta}{2}\sum_k w(\mathbf{k} - \mathbf{q}/2)^2\frac{[\tanh(\beta\xi_\mathbf{k}/2) + \tanh(\beta\xi_\mathbf{k-q}/2)]}{\xi_\mathbf{k} + \xi_\mathbf{k-q} - z} \tag{4.7}$$

where we have also replaced the bosonic Matsubara frequency $i\omega_\nu$ by the complex frequency z. Viewed as a function of z, $A(\mathbf{q}, z)$ has a cut along the real frequency axis for $\Re z \geq -2\mu$ and no other singularity on the (physical) complex plane. In addition, it vanishes for *real* values of z only when $|V|$ is large enough. In this case, we consider the limit $\beta\mu \to -\infty$ and replace Eq. (4.7) by:

$$-\frac{V}{\beta}A(\mathbf{q}, z) = 1 + V\sum_k\frac{w(\mathbf{k})^2}{\frac{k^2}{m} + \frac{q^2}{4m} - 2\mu - z}. \tag{4.8}$$

Let $\omega_\mathbf{q}$ be the solution to the equation

$$A(\mathbf{q}, \omega_\mathbf{q}) = 0 \tag{4.9}$$

for given $\mathbf{q}$. Comparison with Eq. (3.78) (where $2\mu_0 = -\epsilon_0$ is the bound-state energy) yields:

$$\omega_\mathbf{q} = \frac{q^2}{4m} - (2\mu + \epsilon_0). \tag{4.10}$$

It is then clear that the function

$$A'(\mathbf{q}, z) \equiv \frac{A(\mathbf{q}, z)}{\omega_\mathbf{q} - z} \tag{4.11}$$

for given $\mathbf{q}$ is regular also when $z \to \omega_\mathbf{q}$ and non-vanishing over the whole z plane. This remark enables us to rewrite the quadratic action (4.6) in the form

$$S_{eff}^{(2)} = \frac{1}{\beta^2} \sum_q |b(q)|^2 (\omega_\mathbf{q} - i\omega_\nu) A'(\mathbf{q}, i\omega_\nu) , \tag{4.12}$$

which suggests rescaling $b(q)$ by setting

$$b'(q) = \sqrt{A'(\mathbf{q}, i\omega_\nu)} \, \frac{b(q)}{\beta} . \tag{4.13}$$

Expressed in terms of the new variables $b'(q)$, the quadratic action (4.12) reduces to that of a noninteracting Bose system with mass $m_B = 2m$ and chemical potential $\mu_B = 2\mu + \epsilon_0$. Note that the rescaling (4.13) is meaningful insofar as the solution $\omega_\mathbf{q}$ to Eq. (4.9) can be found (i. e., for $|V|$ strong enough that the associated two-body problem possesses a bound state - cf. Eq. (3.78)). In this case, the new field $b'(q)$ acquires the meaning of a truly bosonic field from the $(i\omega_\nu)^{-1}$ decay of its (bare) propagator for large $|\omega_\nu|$ (which, in turn, implies that the correct bosonic commutation rules are recovered for this field).

The rescaling (4.13) obviously affects also the higher $(n > 1)$ terms of the expansion (4.4), which correspond now to the interacting part of the action for the effective bosonic system with "free" action (4.12). In fact, contrary to an ordinary interacting Bose gas for which only the quartic (b^4) interaction exists, bosonization of the original fermionic system has resulted in the infinite set $(b^4, b^6, b^8, \ldots)$ of interactions contained in Eq. (4.4). We shall, however, argue that, in the asymptotic limit of a *dilute* Bose gas obtained from bosonization of the original fermionic system when the condition $\epsilon_F/\epsilon_0 \ll 1$ is satisfied, it is sufficient to retain only the quartic interaction to obtain all physical quantities of interest. It is thus worthed examining first the quartic interaction in some detail.

From Eq. (4.5) we obtain for the term with $n = 2$ of Eq. (4.4):

$$S_{eff}^{(4)} = \frac{1}{4} \mathrm{tr} X^4 = \frac{1}{2} \sum_{q_1 \ldots q_4} I_2(q_1 \ldots q_4) b'(q_1)^* b'(q_2)^* b'(q_3) b'(q_4) \tag{4.14}$$

with

$$I_2(q_1 \ldots q_4) = \delta_{q_1 + q_2, q_3 + q_4} \frac{1}{\sqrt{A'(q_1)^* A'(q_2)^* A'(q_3) A'(q_4)}}$$

$$\times \sum_k w\left(\frac{2k + q_2}{2}\right) w\left(\frac{2k + q_4}{2}\right) w\left(\frac{2k + 2q_2 - q_3}{2}\right) w\left(\frac{2k + 2q_4 - q_1}{2}\right)$$

$$\times \frac{1}{\epsilon(-k)\epsilon(k + q_2)\epsilon(k + q_4)\epsilon(-k + q_1 - q_4)} \tag{4.15}$$

F. Pistolesi

where Eq. (4.13) has been used. Comparison with the standard expression for the quartic interaction [35] then yields:

$$\frac{1}{\beta\Omega}v_2(q_1\ldots q_4) = I_2(q_1\ldots q_4). \tag{4.16}$$

It is clear from Eq. (4.15) that $v_2(q_1\ldots q_4)$ is, in general, a complicated function of its arguments. What is actually relevant for our purposes, however, is knowing (i) the typical "strength" $v_2(0)$ when $q_1 = \ldots = q_4 = 0$ and (ii) the characteristic "range" of its Fourier transform in real space. The latter will be examined in Appendix B in the limit $k_0 \to \infty$ where calculations get considerably simplified. The strength $v_2(0)$ can be evaluated directly from Eq. (4.15) in the limit $\beta\mu \to -\infty$:

$$\begin{aligned} v_2(0) &= \beta\Omega\frac{1}{A'(0)^2}\sum_k w(k)^4\frac{1}{\epsilon(k)^2\epsilon(-k)^2} \\ &= \frac{\beta^2\Omega}{2\pi}\frac{1}{A'(0)^2}\sum_{\mathbf{k}} w(\mathbf{k})^4\int_{-\infty}^{+\infty}d\omega\frac{1}{(\omega^2+\xi_{\mathbf{k}}^2)^2} \\ &= \frac{\beta^2\Omega}{4}\frac{1}{A'(0)^2}\sum_{\mathbf{k}}\frac{w(\mathbf{k})^4}{\xi_{\mathbf{k}}^3} \end{aligned} \tag{4.17}$$

where (cf. Eq. (4.11))

$$A'(0) = \frac{A(0)}{\omega_{\mathbf{q}=0}} = \frac{\beta}{4}\sum_{\mathbf{k}}\frac{w(\mathbf{k})^2}{\xi_{\mathbf{k}}^2} \tag{4.18}$$

at the leading order in the *small* parameter $|\mu_B|/\epsilon_0$. We thus obtain:

$$v_2(0) = 4\frac{\frac{1}{\Omega}\sum_{\mathbf{k}}\frac{w(k)^4}{\xi_{\mathbf{k}}^3}}{\left(\frac{1}{\Omega}\sum_{\mathbf{k}}\frac{w(k)^2}{\xi_{\mathbf{k}}^2}\right)^2}. \tag{4.19}$$

Note that $v_2(0)$ is positive and corresponds to a *repulsive* interaction between the composite bosons. The integrals in Eq. (4.19) can be performed analytically when $\gamma = 1/2$ in Eq. (3.14), yielding (in three dimensions)

$$v_2(0) = \frac{32\pi}{2mk_0}\frac{1}{f(c)} \tag{4.20}$$

with $f(c)$ defined by Eq. (3.82). This result enables us to eliminate $f(c)$ in favor of $v_2(0)$ from Eq. (3.81) and rewrite ξ_{phase}^{BE} in the form

$$(\xi_{phase}^{BE})^2 = \frac{1}{4m_Bn_Bv_2(0)} \tag{4.21}$$

as anticipated in Chapter 3, where $n_B = n/2 = k_F^3/(6\pi^2)$ is the bosonic density. For completeness, we shall verify below that expression (4.21) coincides with the result obtained for a *dilute* Bose gas with repulsive interaction $v(0) = v_2(0)$.

Before proceeding further and considering the remaining interaction terms (with $n > 2$) of Eq. (4.4), it is relevant to discuss the *bosonization condition(s)* which we have exploited in the previous and present Chapters (namely, $\Delta_0 \ll |\mu|$ with $\mu < 0$ - see Ref. 70 - and $|\mu_B| \ll \epsilon_0$, in the order) to reach the BE limit. [The additional condition $|\mu| \ll k_0^2/2m$ introduced in Ref. 70 is instead related to the specific form of the interaction potential (cf. Eqs. (3.13) and (3.14)), which prohibits probing length scales (such as the bound-state radius r_0) smaller than k_0^{-1} (cf. also Ref. 58). From virial theorem it follows, in fact, that $\epsilon_0 \sim r_0^{-2}$, from which $r_0 \gg k_0^{-1}$ can be implemented by requiring $k_0^2/2m \gg \epsilon_0 = 2|\mu|$]. The simplest criterion for dealing with *non-overlapping* composite bosons is

$$r_0 \ll k_F^{-1} \tag{4.22}$$

(k_F^{-1} identifying the average interparticle distance), from which it follows that $\epsilon_F \ll \epsilon_0$. At zero temperature in the broken-symmetry state, criterion (4.22) is equivalent to $\Delta_0 \ll |\mu|$. To show this, we assume that $\Delta_0 \ll |\mu|$ and approximate the (mean-field expression for the) density (in dimensions $d < 4$) as follows:

$$n = \frac{1}{\Omega} \sum_{\mathbf{k}} \left(1 - \frac{\xi_{\mathbf{k}}}{E_{\mathbf{k}}}\right) \cong \frac{\Delta_0^2}{2\Omega} \sum_{\mathbf{k}} \frac{w(\mathbf{k})^2}{\xi_{\mathbf{k}}^2} \sim \Delta_0^2 |\mu|^{d/2-2} \tag{4.23}$$

for sufficiently large k_0. This verifies our assumption consistently, since

$$\frac{\Delta_0^2}{\mu^2} \sim \left(\frac{\epsilon_F}{\epsilon_0}\right)^{d/2} \ll 1. \tag{4.24}$$

At finite temperature, on the other hand, we may use either the Bogolubov result $\mu_B \sim n v_2(0)$ for temperatures lower than the (BE) critical temperature, or the ideal gas value $|\mu_B| \sim n^{2/d}$ for temperatures not too larger than the critical temperature. Taking into account Eq. (4.19), we obtain in the first case:

$$|\mu_B| \sim n|\mu|^{1-d/2} \sim |\mu| \left(\frac{\epsilon_F}{|\mu|}\right)^{d/2} \tag{4.25}$$

that gives

$$\frac{|\mu_B|}{\epsilon_0} \sim \left(\frac{\epsilon_F}{\epsilon_0}\right)^{d/2} \ll 1. \tag{4.26}$$

In the second case we obtain instead:

$$\frac{|\mu_B|}{\epsilon_0} \sim \frac{n^{2/d}}{\epsilon_0} \sim \frac{\epsilon_F}{\epsilon_0} \ll 1. \tag{4.27}$$

Recall that Eqs. (4.8) and (4.17) have been obtained with the condition $\beta\mu \to -\infty$ (which is equivalent to considering temperatures much smaller than the pair dissociation temperature $\sim \epsilon_0$). Implementing the bosonization criterion $\epsilon_F \ll \epsilon_0$ has required us to introduce, in addition, the BE critical temperature ($\ll \epsilon_0$) and to verify the bosonization criterion in distinct temperature regimes.

There remains to verify that the interaction terms with $n > 2$ in Eq. (4.4) can be neglected in comparison with the quartic interaction ($n = 2$), whenever the bosonization condition $\epsilon_F \ll \epsilon_0$ is satisfied. To this end, we write in analogy to Eqs. (4.14)-(4.16):

$$
\begin{aligned}
S_{eff}^{(2n)} &= \frac{1}{2n}\mathrm{tr}X^{2n} \\
&= \frac{(\beta\Omega)^{1-n}}{n}\sum_{q_1\ldots q_{2n}} v_n(q_1\ldots q_{2n})b'(q_1)^*\cdots b'(q_n)^*b'(q_{n+1})\cdots b'(q_{2n}) ,
\end{aligned}
$$

(4.28)

and consider only the case $q_1 = \ldots = q_{2n} = 0$, for which

$$
v_n(0) = (-1)^n \left(\frac{\beta\Omega}{A'(0)}\right)^n \frac{1}{\beta\Omega}\sum_k \frac{w(k)^{2n}}{[\epsilon(k)\epsilon(-k)]^n} .
$$

(4.29)

Note that appropriate powers of $\beta\Omega$ have been introduced in the definition of v_n, in accordance with the standard requirement on the Fourier transform of a generic interaction potential in perturbation theory (no care has, however, been paid to symmetrizing v_n). In this way, the expression (4.29) is finite in the limits $\Omega \to \infty$ and/or $\beta \to \infty$ (cf. Eq. (4.18)). To get an *estimate* of $v_n(0)$, we consider the case $w(k) = 1$ and perform the frequency sum in the limit $\beta\mu \to -\infty$:

$$
\begin{aligned}
v_n(0) &\sim \left(\frac{1}{\Omega}\sum_k \frac{1}{\xi_k^2}\right)^{-n} \frac{1}{\Omega}\sum_k \int_{-\infty}^{+\infty} d\omega \frac{1}{(\omega^2 + \xi_k^2)^n} \\
&\sim \left(\frac{1}{\Omega}\sum_k \frac{1}{\xi_k^2}\right)^{-n} \frac{1}{\Omega}\sum_k \xi_k^{1-2n} \\
&\sim \left(|\mu|^{\frac{d}{2}-2}\right)^{-n} |\mu|^{\frac{d}{2}-2n+1} = |\mu|^{\frac{d}{2}(1-n)+1}
\end{aligned}
$$

(4.30)

in d dimensions [73].

Neglecting the interactions $v_n(0)$ with $n > 2$ with respect to $v_2(0)$ relies on the following argument. Consider the "effective" n-boson interaction L_n depicted in Fig. 4.1, which is assembled from $n > 2$ (bare) bosonic propagators arranged in a loop and n interactions $v_2(0)$:

$$
L_n = (v_2(0))^n \frac{1}{\beta\Omega}\sum_q \frac{1}{(i\omega_\nu - \omega_q)^n}
$$

(4.31)

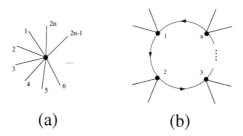

(a)   **(b)**

Figure 4.1: Comparison of (a) the n-body interaction $v_n(0)$ with (b) the "effective" n-boson interaction L_n constructed from $v_2(0)$.

with $w_{\mathbf{q}}$ given by Eq. (4.10) and $\mu_B < 0$. For temperatures not too larger than the (BE) critical temperature (for which Eq. (4.27) holds), it can be readily shown that

$$\frac{1}{\beta\Omega} \sum_q \frac{1}{(i\omega_\nu - w_{\mathbf{q}})^n} \sim \beta^{n-\frac{d}{2}-1} \sim |\mu_B|^{\frac{d}{2}-n+1} \tag{4.32}$$

with a *finite* constant of proportionality. [Note that the result (4.32) could have been guessed directly from dimensional analysis]. Comparison with Eq. (4.30) yields eventually:

$$\frac{v_n(0)}{L_n} \sim \frac{|\mu|^{\frac{d}{2}(1-n)+1}}{\left(|\mu|^{1-\frac{d}{2}}\right)^n |\mu_B|^{\frac{d}{2}-n+1}} \sim \left|\frac{\mu_B}{\epsilon_0}\right|^{n-\frac{d}{2}-1} \ll 1 \tag{4.33}$$

for $n \geq 3$ and $d < 4$, whenever the bosonization criterion $\epsilon_F \ll \epsilon_0$ is satisfied. In this case it is clear that all physical quantities can be obtained by retaining the quartic interaction (v_2) only. Below the critical temperature, on the other hand, the Bogolubov propagator has to be used in Eq. (4.31) in the place of the bare bosonic propagator. This replacement leads to (infrared) divergent integrals, which have to be handled by a suitable renormalization procedure. Although one might argue that an *infinite* constant of proportionality on the right-hand side of Eq. (4.32) would make the condition $v_n(0)/L_n \ll 1$ be satisfied *a fortiori*, consideration of the renormalization procedure is beyond the purposes of the present Part, it will be, instead, the main concern of Part II. This gives a clear warning that a complete description of the crossover from BCS to BE unavoidably requires one to face the peculiar problems arising in the bosonic limit.

In conclusion, we have shown that the action (4.4) can be reduced in the bosonization limit to the simpler form:

$$S_{eff} = \sum_q |b'(q)|^2 (w_{\mathbf{q}} - i\omega_\nu) + \frac{1}{2\beta\Omega} \sum_{q_1 \ldots q_4} v_2(q_1 \ldots q_4) b'(q_1)^* b'(q_2)^* b'(q_3) b'(q_4)$$

$$\tag{4.34}$$

(apart from the constant term $-\mathrm{tr}\,\ln \mathbf{M}_S$), where $v_2(q_1 \ldots q_4)$ depends on its arguments in a complicated way [cf. Eq. (4.15)]. Nonetheless, for many purposes it should be possible to neglect "retardation" effects and replace $v_2(q_1 \ldots q_4)$ with $v_2(0)$ given by Eq. (4.19). In that case, the *mapping* of the original fermionic system *onto a truly bosonic system* is fully established [74].

There remains to recall how the expression (4.21) for ξ_{phase} at zero temperature in the bosonic limit can be obtained *directly* from the bosonic action (4.34) with a constant $v_2(0)$. To this end, we set, as usual, $b'(q) = \sqrt{\beta\Omega}\alpha\delta_{q,0} + \tilde{b}'(q)$ and expand (4.34) up to quadratic order in $\tilde{b}'(q)$:

$$
S_{eff} \cong \beta\Omega|\alpha|^2 \left(\frac{1}{2}v_2(0)|\alpha|^2 - \mu_B \right)
$$

$$
+ \frac{1}{2}\sum_q (\tilde{b}'^{*}(q), \tilde{b}'(-q)) \left(\begin{array}{cc} \frac{\mathbf{q}^2}{2m_B} - i\omega_\nu + |\alpha|^2 v_2(0), & \alpha^2 v_2(0) \\ \alpha^{*2}v_2(0), & \frac{\mathbf{q}^2}{2m_B} + i\omega_\nu + |\alpha|^2 v_2(0) \end{array} \right) \left(\begin{array}{c} \tilde{b}'(q) \\ \tilde{b}'^{*}(-q) \end{array} \right)
$$

$$\tag{4.35}$$

where the Bogolubov self-consistency condition $\mu_B = v_2(0)|\alpha|^2$ has been used. The single-particle bosonic propagators can then be readily obtained by inverting the Gaussian matrix in Eq. (4.35), yielding (in matrix form):

$$
\left\langle \left(\begin{array}{c} \tilde{b}'(q) \\ \tilde{b}'^{*}(-q) \end{array} \right) (\tilde{b}'^{*}(q), \tilde{b}'(-q)) \right\rangle_{S_{eff}}
$$

$$
= \frac{1}{\omega_\nu^2 + E_{\mathbf{q}}^2} \left(\begin{array}{cc} \frac{\mathbf{q}^2}{2m_B} + i\omega_\nu + |\alpha|^2 v_2(0) & -\alpha^2 v_2(0) \\ -\alpha^{*2} v_2(0) & \frac{\mathbf{q}^2}{2m_B} - i\omega_\nu + |\alpha|^2 v_2(0) \end{array} \right) \tag{4.36}
$$

with the Bogolubov quasiparticle dispersion

$$
E_{\mathbf{q}} = \sqrt{(\mathbf{q}^2/2m_B)^2 + 2|\alpha|^2 v_2(0)\mathbf{q}^2/2m_B} \,.
$$

We are actually interested in the longitudinal and transverse correlation functions, defined as [cf. Eqs. (3.5)]

$$
G^{\parallel}(q) = \langle \tilde{b}'_{\parallel}(q)\tilde{b}'_{\parallel}(-q) \rangle \tag{4.37}
$$

$$
G^{\perp}(q) = \langle \tilde{b}'_{\perp}(q)\tilde{b}'_{\perp}(-q) \rangle \tag{4.38}
$$

where [cf. Eqs. (3)]

$$
\tilde{b}'_{\parallel}(q) = \frac{1}{2|\alpha|} \left(\alpha^* \tilde{b}'(q) + \alpha \tilde{b}'^{*}(-q) \right) \tag{4.39}
$$

$$
\tilde{b}'_{\perp}(q) = \frac{1}{2i|\alpha|} \left(\alpha^* \tilde{b}'(q) - \alpha \tilde{b}'^{*}(-q) \right). \tag{4.40}
$$

In terms of the propagators (4.36) we obtain (for real α):

$$G^{\parallel}(q) = \frac{\mathbf{q}^2/(4m_B)}{\omega_{\nu}^2 + E_{\mathbf{q}}^2} \tag{4.41}$$

$$G^{\perp}(q) = \frac{\mathbf{q}^2/(4m_B) + \alpha^2 v_2(0)}{\omega_{\nu}^2 + E_{\mathbf{q}}^2}. \tag{4.42}$$

In particular, the zero-frequency correlation functions read:

$$G^{\parallel}(\mathbf{q}, \omega_{\nu} = 0) = \frac{1}{4m_B}\frac{\mathbf{q}^2}{E_{\mathbf{q}}^2} \cong \frac{1}{4m_B v_S^2}\frac{1}{1 + \mathbf{q}^2 \xi_{phase}^2} \tag{4.43}$$

$$G^{\perp}(\mathbf{q}, \omega_{\nu} = 0) = \frac{\mathbf{q}^2/(4m_B) + \alpha^2 v_2(0)}{E_{\mathbf{q}}^2} \sim \frac{\alpha^2 v_2(0)}{v_S^2 \mathbf{q}^2} \tag{4.44}$$

with the approximate expressions on the right-hand side holding in the small$-\mathbf{q}$ limit, whereby

$$E_{\mathbf{q}} \cong v_S |\mathbf{q}| \sqrt{1 + \mathbf{q}^2 \xi_{phase}^2}. \tag{4.45}$$

Here $v_S = \sqrt{\alpha^2 v_2(0)/m_B}$ is the Bogolubov sound velocity and

$$\xi_{phase} = [4m_B \alpha^2 v_2(0)]^{-\frac{1}{2}}$$

is the desired coherence length for longitudinal correlations. Recalling further that the "condensate" density α^2 coincides with the particle density n_B in the Bogolubov approximation, expression (4.21) (obtained in the bosonization limit) is eventually recovered. This completes our mapping.

In the next Chapter we will show numerically how the crossover for ξ_{phase} progresses from the BCS value (3.71) to the BE value (4.21).

Chapter 5

Numerical Results and Discussion

In Chapter 3 we have identified the coherence length ξ_{phase}, associated with the phase-phase correlation function of a superconducting fermionic system with attractive interaction (3.13), as given by Eq. (3.65) together with Eqs. (3.63) and (3.67). In that Chapter we have also evaluated analytically the asymptotic expressions of ξ_{phase} in the extreme (weak- and strong-coupling) limits. There remains to obtain the behavior of ξ_{phase} in the intermediate-coupling regime, which is especially relevant for the crossover between the two limits. In this regime Eqs. (3.63) and (3.67) have to be evaluated numerically.

To this end, the mean-field parameters μ and Δ_0 need be obtained first. In Appendix C μ and Δ_0 are conveniently expressed in terms of the variable $k_F \xi_{pair}$ in the special case $k_0 = \infty$. A similar scheme can be used for *finite* values of k_0, for which μ and Δ_0 depend *also* on the parameter k_0/k_F. The behavior of μ vs $k_F \xi_{pair}$ for a wide range of values of k_0/k_F has been already given in Ref. 39 and is reported for the sake of comparison in Appendix D for $k_0 = \infty$.

The only mean-field quantity to be discussed here is the expression (3.56) for the longitudinal correlation function. Since the quantity within brackets (with the choice of the minus sign) in Eq. (3.56) coincides with twice the square of the *coherence factor* $(u_{\mathbf{k}} u_{\mathbf{k-q}} - v_{\mathbf{k}} v_{\mathbf{k-q}})$ entering expression (3.61), it is evident by inspection that

$$F_E^{\parallel}(\mathbf{R}) = \frac{V^2}{8} \sum_{\mathbf{q}} e^{i\mathbf{q}\cdot\mathbf{R}} \left(-\frac{1}{V} - f(\mathbf{q}) \right) \approx \frac{V^2}{8} \sum_{\mathbf{q}} e^{i\mathbf{q}\cdot\mathbf{R}} \frac{a'^2}{a' + b\mathbf{q}^2} \tag{5.1}$$

with $f(\mathbf{q})$ given by Eq. (3.64) and

$$a' = \sum_{\mathbf{k}} w(\mathbf{k})^2 \frac{\xi_{\mathbf{k}}^2}{2E_{\mathbf{k}}^3} \ . \tag{5.2}$$

This identifies the characteristic "range" of the function $F_E^{\parallel}(\mathbf{R})$ with $(b/a')^{1/2}$ (cf. Eq. (3.65) and Ref. 68), which is reported vs $k_F \xi_{pair}$ in Fig. 5.1 for the choice $\gamma = 1/2$ in Eq. (3.14). As anticipated in Chapter 3, $F_E^{\parallel}(\mathbf{R})$ can be

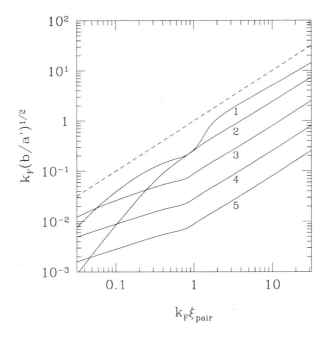

Figure 5.1: Range $(b/a')^{1/2}$ of the function (5.1) (in units of k_F^{-1}) calculated in three dimensions vs $k_F \xi_{pair}$ for $\gamma = 1/2$ and several values of k_0/k_F (defined by 10^{N-1} with $N = 1, 2, \ldots, 5$). The line $k_F \xi_{pair}$ is also shown for comparison (dashed line).

considered to be a "short-range" function of $\mathbf{R}$, since its range never exceeds ξ_{pair} in the BE limit and vanishes when $k_0 \to \infty$. For this reason, the coherence length ξ_{phase} cannot be identified at the mean-field level.

The "long-range" coherence length of interest can be identified instead by the one-loop calculation of Chapter 3 according to Eq. (3.65). In Fig. 5.2 $k_F \xi_{phase}$ is shown vs $k_F \xi_{pair}$ for $\gamma = 1/2$ and several values of k_0/k_F (full lines). Also shown in the figure are: (a) the asymptotic curve (thick line) corresponding to the $k_0 = \infty$ calculation of Appendix C; (b) the boundary (dashed-dotted line) of the "physical" region identified by $c = 1/4$ with c defined after Eq. (3.81); (c) the curve corresponding to $\mu = 0$ (dotted line) where any remnant of the Fermi surface has definitely disappeared; (d) the extrapolation for $k_0 = \infty$ of the analytic BCS and BE results obtained in section 3.3 (dashed line). Note that the values of ξ_{phase} have been uniformly multiplied by the factor $\sqrt{3}/2$ to make ξ_{phase} coinciding with ξ_{pair} in the BCS

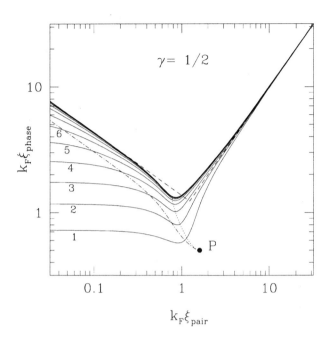

Figure 5.2: $k_F\xi_{phase}$ vs $k_F\xi_{pair}$ for $\gamma = 1/2$ and several values of k_0/k_F (defined by $10^{N/3}$ with $N = 0, 1, \ldots, 10$, such that the "reduced" density is $(k_0/k_F)^{-3} = 10^{-N}$). Values of N label different full curves. Additional conventions are specified in the text.

limit, taking into account their different definitions.

Note also the following features from Fig. 5.2:

(i) ξ_{phase} coincides with ξ_{pair} (*irrespective* of k_0) not only asymptotically in the BCS limit but *also* down to $k_F\xi_{pair} \simeq 10$, where $|\xi_{phase} - \xi_{pair}|/\xi_{pair} < 0.03$ for the values of k_0 reported in the figure.

(ii) For $k_F\xi_{pair} \lesssim 10$ there appears a dependence on k_0, which becomes quite pronounced in the BE limit.

(iii) For given k_0, the minimum value of ξ_{phase} occurs (approximately) at $\mu = 0$ (dotted line).

(iv) The "physical" boundary (dashed-dotted line) and the asymptotic $k_0 = \infty$ curve (thick line) delimit a rather narrow strip for ξ_{phase}.

(v) There exists an accumulation point (denoted by P in the figure) to which the results for $\mu = 0$ converge when $k_0 \to 0$. P belongs also to the "physical" boundary.

(vi) The extrapolation for $k_0 \to \infty$ of the analytic BCS and BE results (dashed line) coincides with the asymptotic $k_0 = \infty$ curve (thick line) *except for* a rather narrow region about $k_F\xi_{pair} \simeq 1$ (or $\mu = 0$). The region where the two curves depart from each other coincides approximately with the "intermediate" region identified in three dimensions from Fig. 6.3 in Appendix D.

Figure 5.2 summarizes the main results of this Chapter. For completeness, we also report in Fig. 5.3 the behavior of $k_F\xi_{phase}$ vs $k_F\xi_{pair}$ using two different

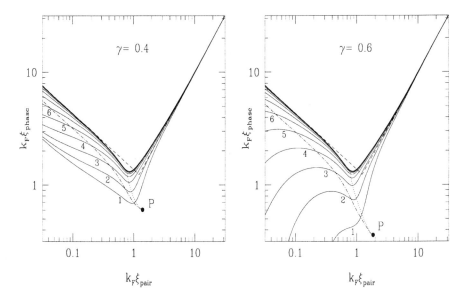

Figure 5.3: $k_F\xi_{phase}$ vs $k_F\xi_{pair}$ for (a) $\gamma = 0.4$ and (b) $\gamma = 0.6$. Conventions are as in Fig. 5.2.

values ($\gamma = 0.4$, 0.6) for the exponent of Eq. (3.14). Note that the conclusions (i) - (vi) drawn above for $\gamma = 1/2$ remain valid, the main difference among results with different values of γ residing in the way they depart from the "physical" boundary. In this sense, the value $\gamma = 1/2$ (considered by NSR) appears to be special.

All results reported above hold specifically in three dimensions. Results of $k_F\xi_{phase}$ vs $k_F\xi_{pair}$ for smaller values of the dimensionality ($2 \leq d \leq 3$) can be obtained by the method of Appendix C in the case $k_0 = \infty$ and are reported in Fig. 5.4. The results for $d = 2$, however, have to be interpreted with caution since fluctuation effects (over and above those considered in the present work) are especially effective in low dimensionality. Note, finally, from Fig. 5.4 that the value $k_F\xi_{pair} = 10$ is still special, since it is (approximately) where the

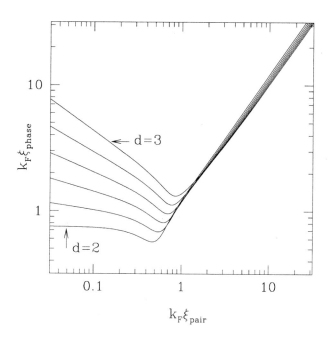

Figure 5.4: $k_F\xi_{phase}$ vs $k_F\xi_{pair}$ for $k_0 = \infty$ and intermediate values of the dimensionality d (in steps of 0.2).

results obtained (with given value of k_0) for different dimensionalities begin to deviate from each other.

Chapter 6

Concluding Remarks

In this Part we have described the zero-temperature behavior of the length ξ_{phase} associated with the fluctuations of the superconducting order parameter, following its crossover from BCS to BE limits. Since the breaking of the gauge symmetry is the phenomenon underlying both superconductivity and superfluidity [75], determining how ξ_{phase} crosses over between the two limits is of definite relevance.

ξ_{phase} has been contrasted with the particle-correlation length ξ_{pair}, which serves to identify the BCS and BE limits as well as to specify the dynamical evolution in between. We have found that ξ_{phase} coincides with ξ_{pair} in the BCS limit and that $\xi_{phase} \gg \xi_{pair}$ in the BE limit, with an interesting behavior in between.

In our calculation we have identified ξ_{phase} and ξ_{pair} using definitions which are valid, in both cases, at the respective significant orders. Specifically, ξ_{pair} has been obtained at the mean-field level and ξ_{phase} at the one-loop order. Some final comments on this procedure, which deals with ξ_{phase} and ξ_{pair} on a different footing, are in order.

ξ_{pair} can be extracted from the fermionic pair-correlation function $g(\mathbf{r})$ defined by Eq. (3.9). Knowledge of ξ_{pair}, in turn, exhausts all relevant information contained in $g(\mathbf{r})$ whenever the underlying dynamical problem possesses a *single* characteristic length. We have verified that this is the case when $g(\mathbf{r})$ is calculated at the mean-field level [cf. Eq. (3.10)], whereby ξ_{pair} reduces to the bound-state radius r_0 in the BE limit. This property, however, might not remain true when fluctuations are included, i.e., by calculating $g(\mathbf{r})$ at the one-loop order via Eq. (3.9). In this case, in fact, a second characteristic length (namely, ξ_{phase}) is expected to appear in $g(\mathbf{r})$. Unfortunately, it is not possible to verify explicitly how ξ_{phase} enters $g(\mathbf{r})$ at the one-loop level by the method of Chapter 3 for evaluating the fermionic correlation functions. It is, in fact, known from the work of Ref. 11 that determining the *density* response function (of which $g(\mathbf{r})$ is a particular case) requires one to include also the coupling between the density and phase-amplitude fluctuations, thus enlarging the Gaussian matrix of Eq. (3.38). Taking into account this coupling exceeds the purposes of the present work. In any event, it should be sufficient to our purposes to identify ξ_{pair} at the mean-field level, since on physical ground no appreciable change is expected for the smallest length scale in the problem when including fluctuations.

We have further argued that ξ_{phase}, on the other hand, cannot be defined at the mean-field level, requiring one to consider explicitly the (one-loop) fluctuation corrections to the longitudinal correlator. It is known, however, that the longitudinal correlator is strongly coupled to the (singular) transverse correlator already at the *next* order of the loop expansion [76]. As a consequence, the longitudinal correlator itself develops singularities for small momenta [77]. Nevertheless, it can be argued that (at least in the bosonic limit) a characteristic length can still be extracted from the longitudinal correlator, this length being identified with our ξ_{phase} [88]. For these reasons, our results for ξ_{phase} vs ξ_{pair} are expected to be essentially correct, both lengths being stable against the inclusion of higher-order fluctuations.

Following our approach to the bosonization problem stated in the Introduction, a deeper understanding of the BE limit should greatly help describing also the crossover problem. As mentioned above, proper treatment of the interacting-boson problem requires special care, owing to the occurrence of infrared singularities that strongly affect the calculation of physical quantities. The study of the large-scale behavior of the bosonic propagator, together with the occurrence of intrinsic infrared singularities, then naturally leads us to consider a renormalization-group approach for handling these singularities. Work along these lines is still required for a full understanding of the interacting-boson problem and Part II is dedicated exactly to this problem. Such an approach, besides being useful for treating the original bosonization problem in the crossover region, has also renewed interest on its own after the recent discovery of a novel Bose-condensed system [78].

Appendix

A Broken-Symmetry Parameter at the One-Loop Order

In this Appendix we prove the identity (3.46), relating the broken-symmetry parameter Δ [cf. Eq. (3.45)] of the original fermionic system to the average of the $q = 0$ component of the bosonic-like variables $b(q)$ introduced via the transformation (3.20). We shall also obtain an explicit expression for the shift Δ_1 of Δ at the one-loop order. Although the explicit value of Δ_1 is irrelevant for the calculation of the phase coherence length of Chapter 3, Δ_1 enters in general the expressions of thermodynamic quantities and correlation functions other than (3.5), for which omitting Δ_1 might lead to inconsistencies. [27,37]

The identity (3.46) is proved by adding to the original fermionic action (3.16) the following bosonic-like source term:

$$\delta S = -J_0 \sum_k w(k) \bar{c}_\uparrow(k) \bar{c}_\downarrow(-k) - J_0^* \sum_k w(k) c_\downarrow(-k) c_\uparrow(k) \qquad (6.1)$$

(with $k = (\mathbf{k}, \omega_s)$). In this way, one obtains from the resulting generating functional analogous to (3.15):

$$\left\langle \sum_k w(k) c_\downarrow(-k) c_\uparrow(k) \right\rangle_S = \left. \frac{\delta Z[\bar{\eta}, \eta; J_0^*, J_0]}{\delta J_0^*} \right|_{\substack{J_0 = J_0^* = 0 \\ \eta = \bar{\eta} = 0}} \qquad (6.2)$$

where the average on the left-hand side is evaluated with the action (3.16). On the other hand, when introducing the Hubbard-Stratonovich transformation (3.20) the additional term (6.1) can be absorbed by shifting the integration variable $b(q)$ with $q = 0$, i.e., by setting $b'(q = 0) = b(q = 0) + \beta J_0$. In this way, one readily obtains:

$$-\frac{1}{V} \langle b(q = 0) \rangle_{S_{eff}} = \left. \frac{\delta Z[\bar{\eta}, \eta; J_0^*, J_0]}{\delta J_0^*} \right|_{\substack{J_0 = J_0^* = 0 \\ \eta = \bar{\eta} = 0}} \qquad (6.3)$$

where now the average on the left-hand side is evaluated with the action (3.28). Comparison of (6.3) with (6.2) yields eventually the result (3.46).

The calculation of the (one-loop) shift Δ_1 is thus equivalent to that of the (one-loop) shift $b_1 = \beta\Delta_1$ of $\langle b(q=0)\rangle_{S_{eff}}$. According to a general procedure of functional integrals (cf., e.g., Appendix C of Ref. 39), this shift is given by

$$b_1 = -e^{i\varphi_0} \frac{\partial F_1/\partial|b_0|}{\partial^2 F_0/\partial|b_0|^2} , \qquad (6.4)$$

where φ_0 is the phase of the source J_0 in Eq. (6.1), F_0 and F_1 are given by Eqs. (3.37) and (3.43), respectively, and $b_0 = \beta\Delta_0$.

Alternatively, Δ_1 can be evaluated in terms of the diagrammatic structure for the original fermionic system via the definition (3.46). This procedure has been used in deriving Eq. (3.57), where the first term on the right-hand side was obtained from the first term on the right-hand side of Eq. (3.52). It is interesting to show explicitly the equivalence of the two procedures at the one-loop order. Besides providing a nontrivial consistency check on our one-loop calculation, the following results may also serve, e.g., to obtain the one-loop correction to the chemical potential over and above its mean-field value.

We begin by writing the single-particle fermionic Green's functions in the form

$$
\begin{aligned}
\langle c_\uparrow(k)\bar{c}_\uparrow(k)\rangle_S &= -\langle \mathbf{M}^{-1}(k,k)_{11}\rangle_{S_{eff}} \\
\langle c_\downarrow(k)\bar{c}_\downarrow(k)\rangle_S &= \langle \mathbf{M}^{-1}(-k,-k)_{22}\rangle_{S_{eff}} \\
\langle c_\uparrow(k)c_\downarrow(-k)\rangle_S &= -\langle \mathbf{M}^{-1}(k,k)_{21}\rangle_{S_{eff}}
\end{aligned}
\qquad (6.5)
$$

where $\mathbf{M}^{-1}$ is the inverse of the matrix (3.25). We can then express the particle density in the form

$$
\begin{aligned}
n &= \frac{1}{\beta\Omega}\sum_{k,\sigma} e^{i\omega_s\delta}\langle \bar{c}_\sigma(k)c_\sigma(k)\rangle_S \\
&= \frac{1}{\beta\Omega}\sum_{k} e^{i\omega_s\delta}\left(\left\langle \mathbf{M}^{-1}(k,k)_{11}\right\rangle_{S_{eff}} - \left\langle \mathbf{M}^{-1}(-k,-k)_{22}\right\rangle_{S_{eff}}\right) \quad (6.6)
\end{aligned}
$$

$(\delta = 0^+)$, and the order parameter Δ in the form

$$\Delta = \frac{V}{\beta}\sum_{k} w(k)\langle c_\uparrow(k)c_\downarrow(-k)\rangle_S = -\frac{V}{\beta}\sum_{k} w(k)\left\langle \mathbf{M}^{-1}(k,k)_{21}\right\rangle_{S_{eff}} . \qquad (6.7)$$

Approximations are introduced at this point in the usual way, by (i) replacing $S_{eff} \to S_{eff}/\lambda$, (ii) implementing the λ-expansion via Eqs. (3.30) and (3.31), and (iii) expanding the resulting expressions in powers of λ. To the first significant order in λ beyond mean field we obtain:

$$
\begin{aligned}
\left\langle \mathbf{M}_\lambda^{-1}(k,k')_{ii'}\right\rangle_{S_{eff}/\lambda} &\cong \delta_{k,k'}\mathbf{M}_0^{-1}(k)_{ii'} + \lambda\left\langle [\mathbf{M}_0^{-1}\mathbf{M}_1\mathbf{M}_0^{-1}\mathbf{M}_1\mathbf{M}_0^{-1}]_{kk'}^{ii'}\right\rangle_{S_{eff}^{(2)}} \\
&\quad -\frac{\lambda}{3}\left\langle [\mathbf{M}_0^{-1}\mathbf{M}_1\mathbf{M}_0^{-1}]_{kk'}^{ii'}\mathrm{tr}(\mathbf{M}_0^{-1}\mathbf{M}_1)^3\right\rangle_{S_{eff}^{(2)}} \quad (6.8)
\end{aligned}
$$

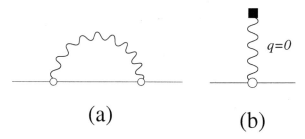

Figure 6.1: (a) Graphical representation of a typical term of Eq. (6.10) (conventions are as in Fig. 3.2); (b) zero-momentum insertion of Eq. (6.11).

where $S^{(2)}_{eff}$ is the (λ-independent) quadratic action (3.38) and the trace is performed over the indices k and i. In deriving Eq. (6.8) we have used the expansion (3.51) for $\mathbf{M}^{-1}_\lambda$ and included consistently the cubic ($n = 3$) term in Eq. (3.34). We shall verify that the first term of order λ on the right-hand side of Eq. (6.8) represents a nontrivial self-energy correction to the bare propagator $\mathbf{M}^{-1}_0$, while the second term of order λ results by shifting the mean-field parameter $\Delta_0 \to \Delta_0 + \Delta_1$ in $\mathbf{M}^{-1}_0$.

There remains to evaluate the contractions entering Eq. (6.8). To this end, it is convenient to supplement the Gaussian action (3.38) by the source term

$$V^{-1}\left(J_0 b'(q=0)^* + J_0^* b'(q=0)\right) \tag{6.9}$$

[which is equivalent to Eq. (6.1)], in order to avoid spurious divergences due to the presence of the Goldstone mode at $q = 0$, allowing J_0 to vanish at the end of the calculation. We obtain eventually:

$$\left\langle [\mathbf{M}^{-1}_0 \mathbf{M}_1 \mathbf{M}^{-1}_0 \mathbf{M}_1 \mathbf{M}^{-1}_0]^{ii'}_{kk'} \right\rangle_{S^{(2)}_{eff}} =$$

$$\delta_{k,k'} \left\{ \mathbf{M}^{-1}_0(k)_{i1} \left[\sum_q w(k+q/2)^2 \mathbf{M}^{-1}_0(k+q)_{21} \left\langle \tilde{b}^*(q)\tilde{b}^*(-q) \right\rangle_{S^{(2)}_{eff}} \right] \mathbf{M}^{-1}_0(k)_{2i'} \right.$$

$$+ \mathbf{M}^{-1}_0(k)_{i1} \left[\sum_q w(k+q/2)^2 \mathbf{M}^{-1}_0(k+q)_{22} \left\langle \tilde{b}(-q)\tilde{b}^*(-q) \right\rangle_{S^{(2)}_{eff}} \right] \mathbf{M}^{-1}_0(k)_{1i'}$$

$$+ \mathbf{M}^{-1}_0(k)_{i2} \left[\sum_q w(k+q/2)^2 \mathbf{M}^{-1}_0(k+q)_{11} \left\langle \tilde{b}(q)\tilde{b}^*(q) \right\rangle_{S^{(2)}_{eff}} \right] \mathbf{M}^{-1}_0(k)_{2i'}$$

$$+ \left. \mathbf{M}^{-1}_0(k)_{i2} \left[\sum_q w(k+q/2)^2 \mathbf{M}^{-1}_0(k+q)_{12} \left\langle \tilde{b}(q)\tilde{b}(-q) \right\rangle_{S^{(2)}_{eff}} \right] \mathbf{M}^{-1}_0(k)_{1i'} \right\}$$

$$\tag{6.10}$$

depicted schematically in Fig. 6.1(a), and

$$-\frac{1}{3} \left\langle [\mathbf{M}^{-1}_0 \mathbf{M}_1 \mathbf{M}^{-1}_0]^{ii'}_{kk'} \mathrm{tr}(\mathbf{M}^{-1}_0 \mathbf{M}_1)^3 \right\rangle_{S^{(2)}_{eff}}$$

$$= \delta_{k,k'} \left\{ \frac{\partial \mathbf{M}_0^{-1}(k)_{ii'}}{\partial b_0^*} b_1^* + \frac{\partial \mathbf{M}_0^{-1}(k)_{ii'}}{\partial b_0} b_1 \right\} \qquad (6.11)$$

depicted in Fig. 6.1(b), with b_1 given by Eq. (6.4).

In particular, when $i = 2$ and $i' = 1$, entering (6.10) and (6.11) into (6.8) with $\lambda = 1$ and the resulting expression into (6.7) yields:

$$\begin{aligned}
\Delta - \Delta_0 &= -\frac{V}{\beta} h + \frac{V}{\beta} \left[B(q = 0)\frac{b_1^*}{\beta} + \left(A(q = 0) + \frac{\beta}{V} \right) \frac{b_1}{\beta} \right] \\
&= \frac{b_1}{\beta} - \frac{V}{\beta} \left[h - (|B(q = 0)| + A(q = 0)) \frac{b_1}{\beta} \right], \qquad (6.12)
\end{aligned}$$

where we have set

$$\begin{aligned}
h &\equiv \sum_k w(k) \sum_q w(k - q/2)^2 \sum_{j,j'=1}^{2} (-1)^{j+j'} \\
&\quad \times \mathbf{M}_0^{-1}(k)_{2j} \mathbf{M}_0^{-1}(q - k)_{j'j} \left\langle \tilde{\mathbf{b}}(q)\tilde{\mathbf{b}}^\dagger(q) \right\rangle_{S_{eff}^{(2)}}^{j'j} \mathbf{M}_0^{-1}(k)_{j'1} \qquad (6.13)
\end{aligned}$$

and made use of Eqs. (3.39) and (3.40). Upon manipulating the derivatives in Eq. (6.4), it can be finally shown that the expression within brackets on the right-hand side of Eq. (6.12) vanishes identically. Equation (6.12) thus reduces to $\Delta - \Delta_0 \equiv \Delta_1 = b_1/\beta$, as expected.

We remark finally that, when $\Delta_0 = 0$, Eq. (6.12) reduces to $A(q = 0)b_1 = 0$, since in this case $h = 0$ identically. This implies that $\Delta_1 = 0$, too.

B   Momentum Dependence of the Interaction Potential for Composite Bosons

In Chapter 4 we have mapped the original fermionic system interacting via an attractive potential onto an effective system of composite bosons, in the limit of strong fermionic attraction. We have also determined the "strength" of the effective *residual* interactions among the composite bosons, which led us to conclude that retaining only the quartic interaction is sufficient to describe the bosonization limit. In this Appendix we study the *momentum* dependence of the quartic interaction, from which we will conclude that ξ_{pair} identifies the characteristic length scale of the boson-boson interaction.

To this end, it is convenient to simplify the expression (4.15) by setting $w(\mathbf{k} = 1)$, that corresponds to taking $k_0 = \infty$ from the outset in Eq. (3.14). It is then clear from dimensional analysis of Eq. (4.15), together with Eqs. (4.8) - (4.11) in the limits $k_0 = \infty$ and $\beta\mu \to -\infty$, that $|\mu|$ and $(2m|\mu|)^{-1/2}$ constitute the only energy and length scales in the problem, respectively. For scattering processes among the composite bosons which involve (Matsubara) frequencies *small* compared to $|\mu|$, one can thus set all (external) bosonic frequencies equal to zero in Eq. (4.15) because to this limit there corresponds a well-defined *finite* value of the interaction potential, as we have verified in Chapter 4. We shall consistently not be particularly interested in the frequency dependence of the effective boson-boson potential. Regarding instead its momentum dependence, we would expect a *truly* bosonic potential to be cast in the (symmetrized) form:

$$v_2(q_1 \ldots q_4) = \delta_{q_1+q_2,q_3+q_4} \left(u(\mathbf{q}_1 - \mathbf{q}_3) + u(\mathbf{q}_1 - \mathbf{q}_4) \right) \tag{6.14}$$

$u(\mathbf{q})$ being the Fourier transform of the two-body interaction potential. In fact, we shall verify below that Eq. (6.14) holds *approximately* only for

$$|\mathbf{q}_i|(2m|\mu|)^{-1/2} \ll 1$$

$(i = 1, \ldots, 4)$ with $u(\mathbf{q}) = $ constant. In other words, the residual boson-boson potential can itself be approximated by a "contact" potential *provided* only small-momentum scattering processes are considered.

To verify to what extent Eq. (6.14) is valid, we consider explicitly two degenerate cases with (i) $q_1 = q_2 = q_3 = q_4 = q$ and (ii) $q_1 = -q_2 = q$ and $q_3 = q_4 = 0$, for which Eq. (6.14) would give $v_2(q, q, q, q) = 2u(\mathbf{q} = 0)$ and $v_2(q, -q, 0, 0) = 2u(\mathbf{q})$, respectively. In the first case, we obtain for the (four) momentum sum in Eq. (4.15) (in three dimensions):

$$\sum_k \frac{1}{\epsilon(k)^2 \epsilon(q-k)^2} = \frac{1}{16\pi} \beta\Omega \frac{(2m|\mu|)^{3/2}}{(2|\mu|)^3} \left[1 + \left(\frac{\mathbf{q}^2}{4m} - i\omega_\nu \right) \frac{1}{2|\mu|} \right]^{-3/2}. \tag{6.15}$$

In the second case we obtain instead:

$$\sum_k \frac{1}{\epsilon(k)\epsilon(-k)\epsilon(k-q)\epsilon(q-k)} = \frac{\beta}{2} \sum_k \frac{1}{\xi_k \left[\xi_{k-q}^2 - (i\omega_\nu - \xi_k)^2 \right]} + \text{c.c.}, \tag{6.16}$$

which for $\omega_\nu = 0$ and in three dimensions reduces to

$$\frac{1}{\pi} \beta\Omega \frac{(2m|\mu|)^{3/2}}{(2|\mu|)^3} \frac{1}{\tilde{q}^2(4+\tilde{q}^2)} \left[(4 + \tilde{q}^2)^{1/2} - 1 - \frac{2}{\pi} \arctan(\tilde{q}/2) - \frac{2}{\pi} \arctan(2/\tilde{q}) \right] \tag{6.17}$$

with $\tilde{q} = |\mathbf{q}|(2m|\mu|)^{-1/2}$. The desired values $v_2(q, q, q, q)$ and $v_2(q, -q, 0, 0)$ are obtained eventually upon dividing the results (6.15) and (6.16), respectively, by $|A'(q)|^2$ and $|A'(q)|A'(0)$.

It is clear from the definition (4.11) of $A'(q)$ (together with (4.8) of $A(q)$) that its computation requires a suitable (ultraviolet) regularization of the momentum integral when $w(\mathbf{k}) = 1$. We follow here a standard procedure in the literature and introduce the scattering amplitude a_s defined via the equation [13, 29]

$$\frac{m}{4\pi a_s} = \frac{1}{\Omega V} + \frac{1}{\Omega} \sum_k \frac{m}{\mathbf{k}^2} \tag{6.18}$$

in the center-of-mass reference frame of the two fermions. The (ultraviolet) divergent sum on the right-hand side of Eq. (6.18) results in a finite value of a_s by letting $V \to 0$ in a suitable way. $A(\mathbf{q}, i\omega_0)$ given by Eq. (4.8) becomes accordingly:

$$\frac{A(\mathbf{q}, i\omega_\nu)}{\beta\Omega} = \frac{1}{\Omega} \sum_\mathbf{k} \left[\frac{m}{k^2} - \left(\frac{k^2}{m} + \frac{q^2}{4m} - 2\mu - i\omega_\nu \right)^{-1} \right] - \frac{m}{4\pi a_s} . \qquad (6.19)$$

Solution of Eq. (4.9), in turn, yields

$$\frac{1}{\Omega} \sum_\mathbf{k} \left[\frac{m}{k^2} - \left(\frac{k^2}{m} + \epsilon_0 \right)^{-1} \right] - \frac{m}{4\pi a_s} = 0 \qquad (6.20)$$

with ϵ_0 defined via Eq. (4.10). In this way we obtain from Eq. (4.11):

$$\begin{aligned}
\frac{A'(\mathbf{q}, i\omega_\nu)}{\beta\Omega} &= \frac{1}{\Omega} \sum_\mathbf{k} \left(\frac{k^2}{m} + \epsilon_0 \right)^{-1} \left(\frac{k^2}{m} + \frac{q^2}{4m} - 2\mu - i\omega_\nu \right)^{-1} \\
&= \frac{1}{4\pi} \frac{m^{3/2}}{\epsilon_0^{1/2}} \frac{\left[1 + \left(\frac{q^2}{4m} - i\omega_\nu \right) \frac{1}{2|\mu|} \right]^{1/2} - 1}{\left(\frac{q^2}{2m} - i\omega_\nu \right) \frac{1}{2|\mu|}}
\end{aligned} \qquad (6.21)$$

where the last equality holds in three dimensions and $2\mu = -\epsilon_0$ within our approximations.

According to Eqs. (4.15) and (4.16), we obtain eventually for $\omega_\nu = 0$ (in three dimensions):

$$\frac{v_2(\mathbf{q}, \mathbf{q}, \mathbf{q}, \mathbf{q})}{v_2(0)} = \frac{1}{2} \frac{\tilde{q}^4}{(4 + \tilde{q}^2)^{3/2}(\sqrt{4 + \tilde{q}^2} - 2)^2} \qquad (6.22)$$

and

$$\frac{v_2(\mathbf{q}, -\mathbf{q}, 0, 0)}{v_2(0)} = \frac{4 \left[(4 + \tilde{q}^2)^{1/2} - 1 - \frac{2}{\pi} \arctan\left(\frac{\tilde{q}}{2} \right) - \frac{2}{\pi} \arctan\left(\frac{2}{\tilde{q}} \right) \right]}{(4 + \tilde{q}^2)(\sqrt{4 + \tilde{q}^2} - 2)} \qquad (6.23)$$

with $\tilde{q}$ defined after Eq. (6.17). The behavior of the expressions (6.22) and (6.23) versus $\tilde{q}$ is depicted in Fig. 6.2. Since $\tilde{q}$ can be also written as the product $|\mathbf{q}|a_s$, from Eqs. (6.22) and (6.23) we conclude that: (i) $v_2(\mathbf{q}, \mathbf{q}, \mathbf{q}, \mathbf{q}) = 2u(\mathbf{q} = 0) = $ constant can be approximately true only for $|\mathbf{q}| \ll a_s^{-1}$ while $v_2(\mathbf{q}, \mathbf{q}, \mathbf{q}, \mathbf{q})$ decays as $(|\mathbf{q}|a_s)^{-1}$ for $|\mathbf{q}| \gg a_s^{-1}$; (ii) the dependence of $v_2(\mathbf{q}, -\mathbf{q}, 0, 0)$ on $|\mathbf{q}|$ is (approximately) given by $v_2(0)4a_s^{-2}/(\mathbf{q}^2 + 4a_s^{-2})$. These results imply that the composite nature of the bosons prevents Eq. (6.14) from holding strictly for "large" momenta (and energies). Nevertheless, our finding that $v_2(\mathbf{q}, -\mathbf{q}, 0, 0)$ decays more rapidly than $v_2(\mathbf{q}, \mathbf{q}, \mathbf{q}, \mathbf{q})$ for $|\mathbf{q}| \gg a_s^{-1}$ makes the assumption (6.14) valid in a "weak" sense.

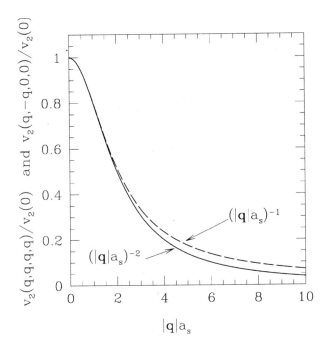

Figure 6.2: Graphical representation of Eqs. (6.22) (dashed line) and (6.23) (full line) vs $\tilde{q} = |\mathbf{q}|a_s$. The characteristic decay of the two functions for large $\tilde{q}$ is indicated.

A final comment is in order. Although we have introduced the scattering amplitude a_s via Eq. (6.18) to comply with a standard procedure in the literature, it is clear that there exists a *single* characteristic length in the bosonic limit, which (in three dimensions) can be identified alternatively with a_s, ξ_{pair}, or r_0 (r_0 being the bound-state radius for the associated two-fermion problem), the three lengths differing at most by numerical constants of order unity. With the "contact" potential $V\Omega\delta(\mathbf{r})$ adopted in this Appendix, in fact, we readily find for the solution of the two-fermion Schrödinger equation in momentum space:

$$\phi(\mathbf{k}) = -\frac{V\sqrt{\Omega}\phi(0)}{\frac{k^2}{m} + \epsilon_0} \tag{6.24}$$

where

$$\phi(0) \equiv \frac{1}{\sqrt{\Omega}}\sum_{\mathbf{k}}\phi(\mathbf{k}) = -V\phi(0)\sum_{\mathbf{k}}\frac{1}{\frac{k^2}{m} + \epsilon_0} \tag{6.25}$$

plays the role of the bound-state equation. Eliminating V in favor of a_s via Eq. (6.18), we find eventually (in three dimensions):

$$\epsilon_0 = \frac{1}{ma_s^2} \ . \tag{6.26}$$

By the same token, the bound-state radius r_0 is given by:

$$r_0^2 = \frac{\sum_{\mathbf{k}} |\nabla_{\mathbf{k}} \phi(\mathbf{k})|^2}{\sum_{\mathbf{k}} |\phi(\mathbf{k})|^2} = \frac{a_s^2}{2} \tag{6.27}$$

where the last equality holds in three dimensions. Since we also know that r_0 coincides with ξ_{pair} in the BE limit, [39] the results of this Appendix could be expressed in terms of ξ_{pair} instead of a_s. This procedure will consistently be adopted in Appendix C also in the BCS regime. Our preference for ξ_{pair} over a_s stands from the fact that ξ_{pair} is (at least in principle) experimentally accessible, since it pertains to the physical problem of interest (while a_s is a fictitious parameter of the theory). Besides, we shall find in Appendix C that expressing the relevant physical quantities in terms of ξ_{pair} from the outset requires no explicit regularization of divergent expressions.

C ξ_{phase} vs ξ_{pair} for a Contact Potential

In the text we have adopted a fermionic interaction potential of the (separable) form (3.13) with $w(\mathbf{k})$ given by Eq. (3.14). By doing so, we have introduced an intrinsic length scale (k_0^{-1}) for the potential, which we have exploited to simplify the regularization procedure and to explore the density dependence of the results. This additional flexibility has enabled us to verify the independence from k_0 of the relevant results in the BCS limit, although physical restrictions limit k_0 to "large" values (cf. Refs. 58 and 70). For this reason we have sometimes considered in the text the limit $k_0 \to \infty$ for the final expressions, where they get considerably simplified (see also Ref. 71).

Purpose of this Appendix is to study *directly* the case $k_0 = \infty$, for which $w(\mathbf{k}) = 1$ and the interaction reduces to a "contact" potential in real space. This potential has already been considered in Appendix B to simplify the calculations; in that context, we have adopted a standard regularization procedure in terms of the scattering amplitude a_s. Here we shall avoid introducing a_s and use ξ_{pair} instead. Setting $k_0 = \infty$ from the outset will also make some approximations used in the text for analytic calculations more transparent.

We begin by evaluating ξ_{pair} according to Eqs. (3.10) and (3.11) in d dimensions:

$$\xi_{pair}^2 = \frac{\sum_{\mathbf{k}} |\nabla_{\mathbf{k}} \phi(\mathbf{k})|^2}{\sum_{\mathbf{k}} |\phi(\mathbf{k})|^2} = \frac{1}{m^2} \frac{\int_0^\infty dk \frac{k^{d+1} \xi_k^2}{E_k^6}}{\int_0^\infty dk \frac{k^{d-1}}{E_k^2}} \tag{6.28}$$

where now $\phi(\mathbf{k}) = 1/E_{\mathbf{k}} = (\xi_{\mathbf{k}}^2 + \Delta_0^2)^{-1/2}$. Here ξ_{pair} is a function of μ and Δ_0 only. Δ_0, in turn, can be related to μ via the (zero-temperature mean-field expression of the) number density (or, alternatively, via k_F):

$$n \equiv \frac{2}{d} K_d k_F^d = K_d \int_0^\infty k^{d-1} \left(1 - \frac{\xi_k}{E_k} \right) \tag{6.29}$$

$(2\pi)^d K_d$ being the area of the unit sphere in d dimensions.

The expression of ξ_{phase} is still given by Eq. (3.65), where now

$$a = \frac{\Delta_0^2 K_d}{2} \int_0^\infty dk \frac{k^{d-1}}{E_k^3} \qquad (6.30)$$

and

$$b = \frac{K_d}{16m} \int_0^\infty dk \frac{k^{d-1}\xi_k^2}{E_k^5} \left[\frac{(\xi_k^2 - 2\Delta_0^2)}{\xi_k} + \frac{5\Delta_0^2}{dm}\frac{k^2}{E_k^2}\right]. \qquad (6.31)$$

Note that the interaction strength V does not appear explicitly in Eqs. (6.28)-(6.31) (this remains true even for $w(\mathbf{k}) \neq$ constant).

Inversion of Eqs. (6.28) and (6.29) yields μ and Δ_0 as functions of k_F and ξ_{pair}, without invoking the gap equation (3.35). The lack of an intrinsic length (such as k_0^{-1}) in the potential enables us to write further:

$$\mu = \mu(k_F, \xi_{pair}) = \frac{k_F^2}{2m} h^{(\mu)}(k_F \xi_{pair}) \qquad (6.32)$$

and

$$\Delta_0 = \Delta_0(k_F, \xi_{pair}) = \frac{k_F^2}{2m} h^{(\Delta_0)}(k_F \xi_{pair}) , \qquad (6.33)$$

where $h^{(\mu)}$ and $h^{(\Delta_0)}$ are functions of the dimensionless variable $k_F \xi_{pair}$ only (this is not true, however, when $w(\mathbf{k}) \neq$ constant). As a consequence, we write from Eqs. (6.30) and (6.31):

$$k_F \xi_{phase} = h^{(\xi)}(k_F \xi_{pair}) \qquad (6.34)$$

where $h^{(\xi)}$ is an additional function of $k_F \xi_{pair}$ only.

Quite generally, Eqs. (6.32)-(6.34) can be solved numerically (in spatial dimensions $d < 4$) for any desired value of $k_F \xi_{pair}$. This procedure has been used in Chapter 5 to determine the limiting curves for $k_0 = \infty$ as well as the dependence of ξ_{phase} on dimensionality. In the rest of this Appendix we discuss the analytic BCS and BE limits.

We consider first the BCS limit and note that the integrals in Eqs. (6.28)-(6.31) can be cast in the form:

$$I_n^m(\mu, \Delta_0) = \int_{-\mu}^{+\infty} d\xi H(\xi) \frac{\xi^m}{E(\xi)^n} , \qquad (6.35)$$

where (m, n) are nonnegative integers and $H(\xi)$ is a smooth function of ξ which, by assumption, does not spoil the ultraviolet convergence of the integral and remains finite for $\xi \to 0$. In the BCS limit, $\Delta_0 \to 0$ and the integral (6.35) develops an infrared singularity when $n - m \geq 1$. Equation (6.35) is then manipulated as follows:

$$I_n^m(\mu, \Delta_0) = H(0) \int_{-\mu}^{+\infty} d\xi \frac{\xi^m}{E(\xi)^n} + \int_{-\mu}^{+\infty} d\xi \left(\frac{H(\xi) - H(0)}{\xi}\right) \frac{\xi^{m+1}}{E(\xi)^n} , \qquad (6.36)$$

where now the first integral on the right-hand side can be evaluated analytically, yielding

$$H(0)\Delta_0^{m-n+1} \int_{-\mu/\Delta_0}^{+\infty} dy \, \frac{y^m}{(y^2+1)^{n/2}} \cong H(0)\Delta_0^{m-n+1} J_n^m \qquad (6.37)$$

with

$$J_n^m \equiv \int_{-\infty}^{+\infty} dy \frac{y^m}{(y^2+1)^{n/2}} \qquad (6.38)$$

since $\mu/\Delta_0 \to \infty$ in the BCS limit. Concerning the second integral on the right-hand side of Eq. (6.36), it may or may not converge in the infrared when $\Delta_0 \to 0$. If it does converge, this term can be safely neglected in comparison to (6.37) for $n - m > 1$; otherwise, the procedure followed in Eq. (6.36) can be iterated for the function $H'(\xi) = (H(\xi) - H(0))/\xi$ in the place of $H(\xi)$, until the resulting integral converges. In any event, the terms generated in this way are sub-leading with respect to (6.37) as $\Delta_0 \to 0$, and can be neglected in the limit. The same procedure has been used in the text to obtain the results (3.69) and (3.70).

With the above approximations, we obtain from the leading terms of Eqs. (6.28), (6.30), and (6.31) in the BCS limit:

$$(\xi_{pair}^{BCS})^2 = \frac{2\mu}{m\Delta_0^2} \frac{J_6^2}{J_2^0} = \frac{\mu}{4m\Delta_0^2} \qquad (6.39)$$

$$(\xi_{phase}^{BCS})^2 = \frac{5\mu}{4dm\Delta_0^2} \frac{J_7^2}{J_3^0} = \frac{\mu}{6dm\Delta_0^2} \qquad (6.40)$$

which recover Eqs. (3.72) and (3.71), respectively, in the limit $k_0 \to \infty$ and $d = 3$. We can thus write in the BCS limit

$$k_F \xi_{phase}^{BCS} = \sqrt{\frac{2}{3d}} k_F \xi_{pair}^{BCS}. \qquad (6.41)$$

In the BE limit, on the other hand, the approximation $\mu/\Delta_0 \to -\infty$ applies and the integrals in (6.28)-(6.31) are conveniently evaluated by expanding their integrands in power of $\Delta_0/|\mu|$. One obtains to leading order:

$$(\xi_{pair}^{BE})^2 = \frac{2}{m|\mu|} \frac{I_1(d)}{I_2(d)}, \qquad (6.42)$$

$$n = \frac{K_d}{2} (2m)^{d/2} \Delta_0^2 \, |\mu|^{\frac{d}{2}-2} I_2(d), \qquad (6.43)$$

$$(\xi_{phase}^{BE})^2 = \frac{|\mu|}{8m\Delta_0^2} \frac{I_2(d)}{I_3(d)}, \qquad (6.44)$$

with $(d < 4)$

$$I_1(d) \equiv \int_0^\infty dy \frac{y^{d+1}}{(y^2+1)^4} = \frac{\Gamma(\frac{d+2}{2})\Gamma(\frac{6-d}{2})}{12} \qquad (6.45)$$

$$I_2(d) \equiv \int_0^\infty dy \frac{y^{d-1}}{(y^2+1)^2} = \frac{\Gamma(\frac{d}{2})\Gamma(\frac{4-d}{2})}{2} \tag{6.46}$$

$$I_3(d) \equiv \int_0^\infty dy \frac{y^{d-1}}{(y^2+1)^3} = \frac{\Gamma(\frac{d}{2})\Gamma(\frac{6-d}{2})}{4}, \tag{6.47}$$

Γ being the Euler's gamma function. Expressing $|\mu|$ in terms of ξ_{pair}^{BE} from (6.42) and Δ_0 in terms of n (and thus of k_F) from (6.43), and entering the results into Eq. (6.44), we obtain in the BE limit:

$$(k_F \xi_{phase}^{BE})^2 = \frac{d}{16} \frac{I_2(d)^2}{I_3(d)} \left(\frac{4 I_1(d)}{I_2(d)} \right)^{\frac{d}{2}-1} (k_F \xi_{pair}^{BE})^{2-d}. \tag{6.48}$$

Note that for $d = 3$ this expression coincides with the result obtained previously in the limit $k_0 \to \infty$ (cf. Ref. 71). Note also that, contrary to the BCS result (6.41) which depends weakly on d, the BE expression (6.48) depends markedly on d and shows a peculiar behavior for $d \to 2$.

D   Behavior of the Chemical Potential vs $k_F \xi_{pair}$

In Chapter 2 (see also Ref. 39) it was found that the crossover between the BCS and BE regimes occurs in a rather *narrow* range of the parameter $k_F \xi_{pair}$, by examining the behavior of the chemical potential vs $k_F \xi_{pair}$ at the mean-field level. This finding has been confirmed by looking at the behavior of $k_F \xi_{phase}$ vs $k_F \xi_{pair}$ with the inclusion of fluctuations. Purpose of this Appendix is to investigate to what extent the behavior of the chemical potential vs $k_F \xi_{pair}$ is "universal", in the sense that it is sufficiently independent from the specific model Hamiltonian and from the dimensionality (at least at the mean-field level).

To this end, we shall examine: (*i*) the continuum model Hamiltonian (3.12)-(3.14) in the limit $k_0 = \infty$ ("contact" potential), for which simplifications occur, at intermediate values of the dimensionality ($2 \leq d \leq 3$); (*ii*) the negative-U Hubbard model on a cubic ($d = 3$) lattice.

The equations determining the chemical potential μ and the gap parameter Δ_0 vs $k_F \xi_{pair}$ for $k_0 = \infty$ and intermediate dimensionality are reported in Appendix C (cf., in particular, Eqs. (6.28) and (6.29)). Their numerical solution yields the behavior of μ vs $k_F \xi_{pair}$ shown in Fig. 6.3 for $2 \leq d \leq 3$. The curve for $d = 3$ coincides with the curve reported in Fig. 1 of Ref. 39 in the limit of low reduced density, and the curve for $d = 2$ coincides with the two-dimensional analytic results given in Ref. 8 (once expressed in terms of $k_F \xi_{pair}$). Note also from Fig. 6.3 that, at the mean-field level, there is no significant difference between the results for $d = 2$ and $d = 3$.

The crossover from BCS to BE for the $d = 3$ negative-U Hubbard model was originally discussed in Ref. 22 in terms of the interaction strength U. Here

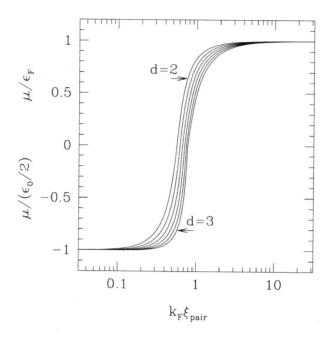

Figure 6.3: Chemical potential μ vs $k_F\xi_{pair}$ (at zero temperature) for $k_0 = \infty$ and $2 \le d \le 3$. Different curves are labeled by the values of d (in steps of 0.2). Positive values of μ are normalized by the Fermi energy $\epsilon_F = k_F^2/2m$, while negative values of μ by half the magnitude ϵ_0 of the eigenvalue of the two-body problem in d dimensions.

we repeat this mean-field calculation, by taking $k_F\xi_{pair}$ (in the place of U) as the variable driving the crossover. The calculation proceeds similarly to that for the continuum model, *but* for the additional inclusion of (normal-state) Hartree-Fock terms in the mean-field decoupling. [65] These terms are now relevant since they provide a sizable shift of the chemical potential near half filling of the electronic band, where they signal the occurrence of a liquid-gas phase separation through a non-monotonic behavior of the chemical potential vs band filling. As the inclusion of pairing restores the correct increase of the chemical potential with filling, a Maxwell construction is required to determine the normal-state value of the chemical potential only.

The chemical potential vs $k_F\xi_{pair}$ for the $d = 3$ negative-U Hubbard model is shown in Fig. 6.4 for several band fillings. Also shown for comparison is the curve for $d = 3$ reproduced from Fig. 6.3 (dotted line). The comparison evidences the peculiar behavior of the Hubbard model near half-filling ($f = 1/2$), while the continuum-model results are recovered in the low-density limit ($f \ll 1$). Note that the qualitative behavior for the Hubbard model looks

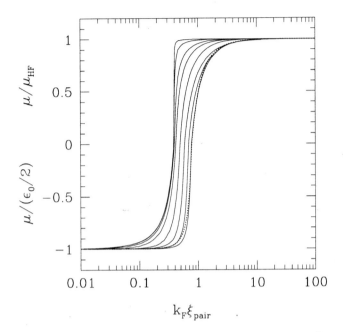

Figure 6.4: Chemical potential vs $k_F\xi_{pair}$ for the $d = 3$ negative-U Hubbard model at the mean-field level. Values of band filling label different curves ($f = 0.01, 0.1, 0.2, 0.3, 0.4, 0.45, 0.49$ from right to left). Energies are measured from the bottom of the single-particle band. The curve with $d = 3$ from Fig. 6.3 is shown for comparison (dotted line). Normalization of μ is as in Fig. 6.3, except for the replacement of the Fermi energy by the Hartree-Fock chemical potential μ_{HF}.

similar to that for the continuum model even at intermediate fillings.

Notwithstanding these similarities, a warning on the nature of the bosonic limit for the negative-U Hubbard model is in order. Contrary to what happens for the continuum model (or else, in the low-density limit $f \ll 1$), the gap equation does not reduce to the bound-state equation for the two-fermion problem when $k_F\xi_{pair} \ll 1$, since near half filling one finds $\Delta = |\mu|$ and the condition $\Delta \ll |\mu|$ cannot be satisfied. As a consequence, the broken-symmetry state is not a BE condensate in the conventional sense, as discussed in Ref. 22.

Part II

IR behavior for $T = 0$ Interacting Bosons

Chapter 7

Introduction to Part II

The problem of interacting bosons in the superfluid phase has been studied extensively in the last fifty years [79–87, 35, 88–93]. Full understanding of this problem has proceeded rather slowly owing to two main difficulties. On the one hand, as Hohenberg and Martin pointed out [85], there is no well-defined prescription to perform consistent approximations to the many-body bosonic problem in the presence of a condensate, since approximations turn out to be either "conserving" or "gapless" (see also the recent Ref. 93). In other words, it is specific to this problem that self-energy approximations yielding a single-particle spectrum without a gap in the limit of vanishing momentum are not consistent with density fluctuations satisfying conservation laws. For this reason, a "small" parameter has been invariably introduced to control the approximations, as for the low-density or weak-coupling limits [79, 80, 35] and for small momentum $\mathbf{k}$ and frequency ω [82]. On the other hand, irrespective of the approximation scheme, infrared (IR) divergences have unavoidably plagued the theory owing to the presence of the Goldstone mode [80, 82, 86, 35, 90–92]. Although most physical quantities turn out to be free of divergences [80], in general it is not possible to control the approximations through a "small" parameter owing to the occurrence of divergences at intermediate steps of the calculation. In particular, the lowest-order Bogolubov result of a non-vanishing anomalous self-energy ($\Sigma_{12}(\omega = 0, \mathbf{k} = 0)$) at zero external momentum and frequency was recognized to be incorrect when its vanishing was established as an exact result [86]. Similarly, IR divergences have posed problems also to the treatment given by Gavoret and Nozières, who introduced an artificial gap Δ in the phonon spectrum to regularize their theory but realized that exchanging the limits $(\mathbf{k}, \omega) \to 0$ and $\Delta \to 0$ could be ill-defined [82].

Nepomnyashchii and Nepomnyashchii [86] (from here on referred to as NN) where the first ones to realize that the Bogolubov approximation is actually incorrect insofar as it relies on $\Sigma_{12}(\omega = 0, \mathbf{k} = 0)$ being non-vanishing. Their finding that $\Sigma_{12}(\omega = 0, \mathbf{k} = 0)$ vanishes exactly would then apparently imply the disappearance of the sound mode and hence of superfluidity itself.

Nonetheless, NN found that the sound mode was preserved even when the correct result $\Sigma_{12}(\omega = 0, \mathbf{k} = 0) = 0$ was taken into account. The NN results were found by studying directly the skeleton structure of the diagrammatic perturbation theory (PT). To determine the asymptotic behavior of the correlation functions for $k \to 0$, however, NN had to resort to a brute-force cancellation of IR divergences which, besides being not fully transparent, might be not readily controlled.

The need of developing a PT free from IR divergences was actually realized by Popov [35] before NN. This was achieved within a functional-integral approach by introducing a phase-amplitude representation for the bosonic variables in the IR region for $|\mathbf{k}|$ smaller than a cutoff parameter k_o. In this way all intermediate steps of the calculations turned out to be free from IR divergences, allowing the approximations to be controlled by the use of a "small" parameter. However, elimination of the arbitrary parameter k_o remains nontrivial in the Popov approach, as it requires full control of the IR region up to momenta of the order of k_o where the phase-amplitude description becomes less and less accurate, since at large momenta the particle representation is appropriate. In addition, the phase-amplitude description somewhat obscures the particle viewpoint even at small momenta. As a matter of fact, when $k \to 0$ it is not easy to follow with this description the *nontrivial* evolution of the elementary excitations from free particles to the sound mode. Probably for this reason Popov recognized in his own language the NN result about the vanishing of the anomalous self-energy only at a later stage. [87]

The vanishing of the anomalous self-energy is not a mere mathematical result but it has a definite *physical* origin, as it was lately recognized by NN. In this particular context of broken gauge symmetry, this finding reflects the quite general picture given by Patašinskij and Pokrovskij [94], whereby the Goldstone-mode singularity of the transverse correlation function drives a divergence in the longitudinal correlation function for a continuous broken symmetry.

All these circumstances lead to the necessity of providing a unifying and fully controlled treatment of IR divergences. In this respect, the renormalization group (RG) approach appears to be the natural and reliable tool to determine the IR behavior of the system in the presence of IR divergences. In this Part we exploit the RG approach to obtain the *exact* IR behavior of all vertex functions (and thus of all correlation functions) for a neutral system of interacting bosons in the broken-symmetry phase at zero temperature. To this end, we will make extensive use of the Ward identities (WI) associated with the gauge invariance, which pose strong conditions on the RG equations and enable us to obtain the desired solution to all orders of the ϵ-expansion. Although Popov method might at first look optimally suited to this purpose since it concentrates from the outset on variables which are manifestly gauge invariant, we believe that it is useful to complement the phase-amplitude approach by studying the IR behavior of the system via the more standard particle representation.

Usually, the appearance of IR divergences is related to a second-order phase transition, when a competition between two different phases with equal free energy leads to divergent fluctuations at the critical temperature. In the present case, although the system is in a stable phase and no phase transition occurs, there exists a competition among degenerate states which are associated with different values of the macroscopic phase of the order parameter and are thus physically equivalent. This leads to the divergence of the longitudinal correlation function mentioned above, while all correlation functions obtained as averages of local gauge-invariant operators are expected to be free from IR divergence (like the density-density correlation function).

Perturbation theory for the broken symmetry phase in the particle representation considers averages of operators which are not local gauge invariant and is therefore plagued by IR divergences at intermediate steps of the calculation. It is clear that these divergences cannot be independent from each other as they have to cancel out when calculating averages of the above-mentioned local gauge-invariant operators. In this respect, WI provide explicit connections among the divergences. In turn, this makes *all* gauge-invariant susceptibilities (like the compressibilities) finite and accordingly stabilizes the system with respect to phase fluctuations.

By our RG approach we will be able to prove that the three available gauge-invariant susceptibilities of the system (namely, the condensate susceptibility, the ordinary susceptibility - which is related to the sound velocity - and the superfluid density - which is related to the transverse current-current response function) are indeed finite, being invariants of the RG flow. This will be done by establishing explicit connections among the running couplings. In this way we will be left with only one independent running coupling, whose infrared behavior will then be established to *all* orders in ϵ $(= 3-d)$ for $d > 1$. This will result into a novel line of fixed points different from the usual Bogolubov line of fixed points. Nonetheless, the occurrence of the sound spectrum will turn out to be independent from the scaling of the independent running couplings, as it depends only on the underlying gauge symmetry. We will also recover the leading IR behavior of the single-particle Green's functions and of the (two-particle) response functions given previously by Gavoret and Nozières [82] and by NN [86]. Finally, we will provide an independent justification of the fact that the ϵ-expansion can be safely extended down to $d = 1^+$, by suitably reorganizing the skeleton structure of the diagrammatic PT using the RG procedure as a guide. In this way the mechanism of how the divergences sum up to give the correct IR behavior for all vertex functions will be greatly clarified.

Previous RG treatment of the zero-temperature interacting-boson problem have been given by Weichman [88] and by Benfatto [92]. Weichman found the correct exponent of the longitudinal susceptibility by performing a one-loop calculation for an intrinsically non-divergent quantity (the free energy) and then used a simple scaling argument to determine the singular behavior of the longitudinal single-particle Green's function. This scaling argument can, in general, be justified by the RG approach, as it will be shown below by our treatment. Benfatto, on the other hand, used a Wilson-like RG approach in a rigorous fashion to determine the scaling behavior for a number of running couplings. However, his treatment was limited to $d = 3$ and lacked a systematic implementation of WI, which in turn would guarantee the linear behavior of the spectrum irrespective of the IR behavior of the longitudinal correlation function. We will find that the asymptotic freedom found by Benfatto for $d = 3$ is not essential to solve exactly the IR problem and to stabilize the superfluid phase by preserving the linear spectrum.

The question naturally arises whether the momentum region, where the phase fluctuations (leading to IR divergences) dominate, has physical relevance apart from establishing the superfluid behavior. To this end, a generalized Ginzburg criterion can be introduced on the momentum variable to determine, in particular, whether the IR region will merge or not into the Bogolubov region, $i.e.$, where the spectrum is still linear but the longitudinal correlation function is approximately constant. For the low-density Bose gas one can explicitly show [87] that the Ginzburg region extends up to a characteristic momentum that vanishes exponentially (in the gas parameter) with respect to the extension of the linear Bogolubov region. In the more general case it is not a $priori$ possible to assess whether the Bogolubov region might be washed out by the growing of the Ginzburg region. In this case a full RG approach is required.

The plan of Part II is the following. In Chapter 8 we introduce the action and write various Ward identities that will be used in the rest of the part. In Chapter 9 we classify the vertex functions on the basis of their relevance. We thus introduce the perturbation theory and find constraints given by the symmetry on the correlations functions. In Chapter 10 we use the Ward Identities to connect the different running couplings, and in particular we fix their limiting values. In Chapter 11 the RG approach is used to study the IR behavior of the system; the β-functions are evaluated at the order of one-loop and the generalization of the results to all order is discussed. In this way an explicit expression for the single-particle Green's functions and the response functions is found and, as an example, the depletion of the condensate is studied within the RG approach. In Chapter 12 we give the conclusions and perspectives extension of this Part.

Chapter 8

Bose Action and Ward Identities

8.1 The Action in Presence of Condensate

To describe the (short range) interacting Bose system we consider the following action:

$$S = S_Q + S_{int} + S_\lambda \qquad (8.1)$$

where ($\hbar = 1$, $m = 1/2$)

$$
\begin{aligned}
S_Q &= \int dx \left\{ \psi^*(x^+) \left[-\partial_\tau + \mu(x) \right] \psi(x) \right. \\
&\quad \left. - (\nabla + i\mathbf{A}(x)) \psi^*(x^+) (\nabla - i\mathbf{A}(x)) \psi(x) \right\},
\end{aligned}
\qquad (8.2)
$$

$$S_{int} = -\frac{1}{2} v \int dx \psi^{*2}(x) \psi^2(x), \qquad (8.3)$$

and

$$S_\lambda = \frac{1}{2} \int dx \left[\psi(x) \lambda^*(x) + \psi^*(x) \lambda(x) \right]. \qquad (8.4)$$

$x = (\tau, \mathbf{x})$ and $\psi(\tau, \mathbf{x})$ is a periodic function of τ (with period β) and $\mathbf{x}$ (with period L) [35]. The integral is performed over the following multidimensional rectangle:

$$
\begin{aligned}
\int dx &= \int_0^\beta d\tau \int_0^L dx_1 \ldots \int_0^L dx_d \\
&= \int_{-\beta/2}^{\beta/2} d\tau \int_{-L/2}^{L/2} dx_1 \ldots \int_{-L/2}^{L/2} dx_d.
\end{aligned}
\qquad (8.5)
$$

In (8.1) we have introduced the external sources λ and $\mathbf{A}$ to generate the connected correlation functions of the theory. In this way the sources will be let eventually to vanish. The $\mu(x)$ source will instead reduce to the uniform chemical potential μ of the system.

The action (8.1) allows for spontaneous symmetry breaking; in that case, it is convenient to distinguish between longitudinal χ_1 and transverse χ_2 components to the broken-symmetry direction [92]. We thus set:

$$\psi(x) = \chi_1(x) + i\chi_2(x) \tag{8.6}$$
$$\psi^*(x) = \chi_1(x) - i\chi_2(x), \tag{8.7}$$

with $\chi_i(x)$ real functions, and the measure of the Functional integral becomes:

$$\mathcal{D}\psi(x)^*\mathcal{D}\psi(x) = \mathcal{D}\chi_1(x)\mathcal{D}\chi_2(x). \tag{8.8}$$

In terms of the new fields we are left with the following Action:

$$S = \int dx \left\{ -i\chi_1\partial_\tau\chi_2 + i\chi_2\partial_\tau\chi_1 + (\mu - \mathbf{A}^2)(\chi_1^2 + \chi_2^2) - (\nabla\chi_1)^2 - (\nabla\chi_2)^2 \right.$$
$$\left. -2\mathbf{A}\chi_2\nabla\chi_1 + 2\mathbf{A}\chi_1\nabla\chi_2 - \frac{v}{2}(\chi_1^4 + \chi_2^4) - v\chi_1^2\chi_2^2 + \lambda_1\chi_1 + \lambda_2\chi_2 \right\}. \tag{8.9}$$

We can perform now an additional change of variables corresponding to breaking the Gauge symmetry [95]:

$$\chi_1(x) = \chi_{1o}(x) + \tilde{\chi}_1(x) \tag{8.10}$$
$$\chi_2(x) = \tilde{\chi}_2(x). \tag{8.11}$$

In this way we obtain the following expression for the action:

$$S[\tilde{\chi}, A_\nu, \lambda_i] =$$
$$= \int dx \left\{ (\mu - \mathbf{A}^2)\chi_{1o}^2 - (\nabla\chi_{1o})^2 + \lambda_1\chi_{1o} - \frac{v}{2}\chi_{1o}^4 \right.$$
$$+ \left[2(\mu - \mathbf{A}^2)\chi_{1o} - 2(\nabla\chi_{1o})\nabla - 2v\chi_{1o}^3 + \lambda_1 \right]\tilde{\chi}_1$$
$$+ \left[-i\chi_{1o}\partial_\tau + i\partial_\tau\chi_{1o} - 2\mathbf{A}(\nabla\chi_{1o}) + 2\chi_{1o}\mathbf{A}\nabla + \lambda_2 \right]\tilde{\chi}_2$$
$$+ \left[\mu - \mathbf{A}^2 - 3v\chi_{1o}^2 \right]\tilde{\chi}_1^2 - (\nabla\tilde{\chi}_1)^2$$
$$+ \left[\mu - \mathbf{A}^2 - v\chi_{1o}^2 \right]\tilde{\chi}_2^2 - (\nabla\tilde{\chi}_2)^2$$
$$-i\tilde{\chi}_1\partial_\tau\tilde{\chi}_2 + i\tilde{\chi}_2\partial_\tau\tilde{\chi}_1 + 2\mathbf{A}\tilde{\chi}_1\nabla\tilde{\chi}_2 - 2\mathbf{A}\tilde{\chi}_2\nabla\tilde{\chi}_1$$
$$\left. -2v\chi_{1o}\tilde{\chi}_1^3 - 2v\chi_{1o}\tilde{\chi}_1\tilde{\chi}_2^2 - \frac{v}{2}\left[\tilde{\chi}_1^4 + \tilde{\chi}_2^4 + 2\tilde{\chi}_1^2\tilde{\chi}_2^2 \right] \right\}. \tag{8.12}$$

If the external sources λ and $\mathbf{A}$ vanish, we have a uniform condensate $\chi_{1o}(x) = \chi_{1o}$ and the expression for the action greatly simplifies:

$$S[\tilde{\chi}, \mu] =$$
$$= \int dx \left\{ \mu\chi_{1o}^2 - \frac{v}{2}\chi_{1o}^4 + \left[2\mu\chi_{1o} - 2v\chi_{1o}^3 \right]\tilde{\chi}_1 \right.$$
$$+ \left[\mu - 3v\chi_{1o}^2 \right]\tilde{\chi}_1^2 - (\nabla\tilde{\chi}_1)^2 + \left[\mu - v\chi_{1o}^2 \right]\tilde{\chi}_2^2 - (\nabla\tilde{\chi}_2)^2 - i\tilde{\chi}_1\partial_\tau\tilde{\chi}_2 + i\tilde{\chi}_2\partial_\tau\tilde{\chi}_1$$
$$\left. -2v\chi_{1o}\tilde{\chi}_1^3 - 2v\chi_{1o}\tilde{\chi}_1\tilde{\chi}_2^2 - \frac{v}{2}\left[\tilde{\chi}_1^4 + \tilde{\chi}_2^4 + 2\tilde{\chi}_1^2\tilde{\chi}_2^2 \right] \right\}. \tag{8.13}$$

We define now the Fourier transform of the fields:

$$\tilde{\chi}_i(\omega_\nu, \mathbf{k}) = \frac{1}{(\beta\Omega)^{1/2}} \int_0^\beta d\tau \int d^d x \, \tilde{\chi}_i(x) e^{-ikx} \tag{8.14}$$

$$\tilde{\chi}_i(\tau, x) = \frac{1}{(\beta\Omega)^{1/2}} \sum_{\omega_\nu, \mathbf{k}} \tilde{\chi}_i(k) e^{ikx}.. \tag{8.15}$$

The condition of reality for the $\tilde{\chi}_i$ fields leads to the usual relation between the Fourier components:

$$\tilde{\chi}_i(x)^* = \tilde{\chi}_i(x) \quad \Rightarrow \quad \tilde{\chi}_i(k) = \tilde{\chi}_i(-k)^*, \tag{8.16}$$

this relation states that the fields $\tilde{\chi}_i(k)$ are not independent. In terms of the Fourier components the action becomes:

$$
\begin{aligned}
S[\tilde{\chi}, \mu] = \\
= \ & \beta\Omega \left[\mu \chi_{1o}^2 - \frac{v}{2} \chi_{1o}^4 \right] + (\beta\Omega)^{1/2} \left[2\mu\chi_{1o} - 2v\chi_{1o}^3 \right] \tilde{\chi}_1(0) \\
& + \sum_k \left\{ \left[\mu - 3v\chi_{1o}^2 - \mathbf{k}^2 \right] \tilde{\chi}_1(k)\tilde{\chi}_1(-k) + \left[\mu - v\chi_{1o}^2 - \mathbf{k}^2 \right] \tilde{\chi}_2(k)\tilde{\chi}_2(-k) \right. \\
& \left. + 2\omega_\nu \tilde{\chi}_1(-k)\tilde{\chi}_2(k) \right\} \\
& - 2v\chi_{1o}(\beta\Omega)^{-1/2} \sum_{k_1+k_2+k_3=0} \left\{ \tilde{\chi}_1(k_1)\tilde{\chi}_1(k_2)\tilde{\chi}_1(k_3) + \tilde{\chi}_1(k_1)\tilde{\chi}_2(k_2)\tilde{\chi}_2(k_3) \right\} \\
& - (\beta\Omega)^{-1} \sum_{k_1+k_2+k_3+k_4=0} \frac{v}{2} \left\{ \tilde{\chi}_1(k_1)\tilde{\chi}_1(k_2)\tilde{\chi}_1(k_3)\tilde{\chi}_1(k_4) \right. \\
& \left. + \tilde{\chi}_2(k_1)\tilde{\chi}_2(k_2)\tilde{\chi}_2(k_3)\tilde{\chi}_2(k_4) + 2\tilde{\chi}_1(k_1)\tilde{\chi}_1(k_2)\tilde{\chi}_2(k_3)\tilde{\chi}_2(k_4) \right\} \tag{8.17}
\end{aligned}
$$

811   MeanField Propagators

The quadratic part of (8.17) can be written in matrix form as follows:

$$S_Q = -\frac{1}{2} \sum_k \tilde{\chi}_i(-k) M_{ij}(k) \tilde{\chi}_j(k), \tag{8.18}$$

where

$$M(k) = -2 \begin{pmatrix} \mu - 3v\chi_{1o}^2 - \mathbf{k}^2 & \omega_\nu \\ -\omega_\nu & \mu - v\chi_{1o}^2 - \mathbf{k}^2 \end{pmatrix}. \tag{8.19}$$

We define the propagator $\mathcal{G}_{ij}$:

$$\mathcal{G}_{ij}(k) \equiv \langle \tilde{\chi}_i(k)\tilde{\chi}_j(-k) \rangle \equiv \frac{\int \mathcal{D}\tilde{\chi}_1 \mathcal{D}\tilde{\chi}_2 \, e^S \tilde{\chi}_i(k)\tilde{\chi}_j(-k)}{\int \mathcal{D}\tilde{\chi}_1 \mathcal{D}\tilde{\chi}_2 e^S}, \tag{8.20}$$

at the mean field level $\mathcal{G}^o$ is given simply by $M(k)^{-1}$ [96]:

$$\mathcal{G}^o(k) = -\frac{1}{2} \frac{\begin{pmatrix} \mu - v\chi_{1o}^2 - \mathbf{k}^2 & -\omega_\nu \\ \omega_\nu, & \mu - 3v\chi_{1o}^2 - \mathbf{k}^2 \end{pmatrix}}{(\mathbf{k}^2 - \mu + 3v\chi_{1o}^2)(\mathbf{k}^2 - \mu + v\chi_{1o}^2) + \omega_\nu^2} \tag{8.21}$$

If we choose the free parameter χ_{1o} as the value that eliminates the linear term in the action (8.17) we have the mean field condition:

$$\chi_{1o}^2 = \mu/v \qquad (8.22)$$

and the mean field propagator becomes:

$$\mathcal{G}^o(k) = \frac{1}{2} \frac{1}{(\mathbf{k}^2 + 2\mu)\mathbf{k}^2 + \omega_\nu^2} \begin{pmatrix} \mathbf{k}^2 & \omega_\nu \\ -\omega_\nu, & 2\mu + \mathbf{k}^2 \end{pmatrix} \qquad (8.23)$$

so for $\mathbf{k}$ and ω small (for ω_ν we refer of course to the analytic continuation) the leading IR behavior for the three propagators is the following:

$$\mathcal{G}_{11}(k) \quad \sim \quad \mathbf{k}^2/(\omega^2 + c_o^2\mathbf{k}^2) \qquad (8.24)$$
$$\mathcal{G}_{12}(k) \quad \sim \quad \omega/(\omega^2 + c_o^2\mathbf{k}^2) \qquad (8.25)$$
$$\mathcal{G}_{22}(k) \quad \sim \quad c_o^2/(\omega^2 + c_o^2\mathbf{k}^2) \qquad (8.26)$$

where we have defined the Bogolubov mean field sound velocity: $c_o^2 = 2\mu$. In the present context we are interested in the IR behavior of the correlation function, *i.e.*, in the following limit:

$$\lim_{\alpha \to 0} \mathcal{G}(\alpha\omega, \alpha\mathbf{k}) . \qquad (8.27)$$

It is important at this point to recognize the advantage of using transverse and longitudinal fields. In fact, with this choice we have kept the strongest Goldstone singularity in the transverse propagator $\mathcal{G}_{22}(k)$. This must be contrasted with the usual ψ-representation in which all Bogolubov propagators share the same IR behavior. This choice will result crucial to select the interaction terms on the basis of their relevance.

8.2   Gauge Invariance and Ward Identities

The Bosonic action introduced in (8.1) is invariant under the following Gauge transformation ($\alpha(x)$ satisfy periodic boundary conditions):

$$\begin{cases} \psi(x) & \to & \psi(x)e^{i\alpha(x)} \\ \psi(x)^* & \to & \psi(x)^*e^{-i\alpha(x)} \\ \mathbf{A}(x) & \to & \mathbf{A}(x) + \nabla\alpha(x) \\ \mu(x) & \to & \mu(x) + i\partial_\tau\alpha(x) \end{cases} \qquad (8.28)$$

In terms of our fields the same invariance reads:

$$S[T_{ij}\chi_j(x), A_\nu + \partial_\nu\alpha(x), T_{ij}\lambda_j] = S[\chi_i(x), A_\nu, \lambda_i], \qquad (8.29)$$

where $A_\nu = (\mu(x), \mathbf{A}(x))$, $\partial_\nu = (i\partial_\tau, \nabla)$, (we define also $k_\nu = (i\omega, \mathbf{k})$) and:

$$T = \begin{pmatrix} \cos\alpha(x) & -\sin\alpha(x) \\ \sin\alpha(x) & \cos\alpha(x) \end{pmatrix}. \tag{8.30}$$

We define the Free energy F with the following expression:

$$F[A_\nu, \lambda_i] = \ln\left\{\int \mathcal{D}\chi^* \mathcal{D}\chi \exp\{S[\chi, A_\nu, \lambda_i]\}\right\}, \tag{8.31}$$

In this way the invariance of S will result in the following invariance property of F:

$$F[A_\nu + \partial_\nu \alpha(x), T_{ij}\lambda_j] = F[A_\nu, \lambda_i]. \tag{8.32}$$

We can proceed by defining the Legendre transform of F:

$$\chi_{io}(x) = \frac{\delta F}{\delta \lambda_i(x)} = \langle \chi_i(x) \rangle. \tag{8.33}$$

We invert formally the last equation to obtain λ_i as a functional of χ_{io}, in this way we define Γ:

$$\Gamma[\chi_{io}(x), A_\nu(x)] = \int dx \lambda_i(x)\chi_{io}(x) - F[A_\nu, \lambda_i[\chi_{io}]]. \tag{8.34}$$

Useful properties of Γ are the following:

$$\frac{\delta\Gamma}{\delta\chi_{io}(x)} = \lambda_i(x), \tag{8.35}$$

$$\frac{\delta\Gamma}{\delta A_\nu(x)} = -\frac{\delta F}{\delta A_\nu(x)}. \tag{8.36}$$

We now define the functional derivatives of Γ:

$$\frac{\delta^{(n)}\Gamma}{\delta\chi_{i_1 o}(x_1)\dots\delta\chi_{i_n o}(x_n)} = \Gamma_{i_1\dots i_n}(x_1,\dots,x_n) \tag{8.37}$$

$$\frac{\delta^{(n)}\Gamma}{\delta A_{\nu_1}(y_1)\dots\delta A_{\nu_n}(y_n)} = \Gamma_{;\nu_1\dots\nu_n}(y_1,\dots,y_n) \tag{8.38}$$

and in general the mixed derivatives:

$$\frac{\delta^{(n+m)}\Gamma}{\delta\chi_{i_1 o}(x_1)\dots\delta\chi_{i_n o}(x_n)\delta A_{\nu_1}(y_1)\dots\delta A_{\nu_n}(y_m)}$$
$$= \Gamma_{i_1\dots i_n;\nu_1\dots\nu_m}(x_1,\dots,x_n;y_1,\dots,y_m). \tag{8.39}$$

From the definition of the correlation function we can see that the derivatives of Γ are the vertex functions of the theory. In fact, by differentiating Eq. (8.35) with respect to $\lambda_j(y)$ we obtain:

$$\int dz \Gamma_{ii'}(x,z)\mathcal{G}_{i'j}(z,y) = \delta_{ij}\delta(x-y), \tag{8.40}$$

where

$$\mathcal{G}_{ij}(x,y) = \langle \chi_i(x)\chi_j(y)\rangle - \langle \chi_i(x)\rangle \langle \chi_j(y)\rangle = \frac{\delta^{(2)}F}{\delta\lambda_i(x)\delta\lambda_j(y)}. \qquad (8.41)$$

We define the Connected Green's function in the same way we have defined the vertex parts:

$$\frac{\delta^{(n+m)}F}{\delta\lambda_{i_1}(x_1)\dots\delta\lambda_{i_n}(x_n)\delta A_{\nu_1}(y_1)\dots\delta A_{\nu_n}(y_m)} = \mathcal{G}_{i_1\dots i_n;\nu_1\dots\nu_m}(x_1,\dots,x_n;y_1,\dots,y_n).$$
$$(8.42)$$

We can continue differentiating (8.40); this will lead to an infinite series of equations that link the vertex parts with the Green's functions. From these equations one can prove the irreducibility with respect to the cut of one particle lines of the vertex part (see in particular Sec. 9.2.3).

The local gauge invariance of S leads, in a similar way, to the following invariance for Γ:

$$\Gamma[A_\nu + \partial_\nu\alpha(x), T_{ij}\chi_{jo}] = \Gamma[A_\nu, \chi_{io}]. \qquad (8.43)$$

821   Ward Identities

Now we consider the consequences of the Gauge invariance on the vertex parts of the theory. We begin by functional differentiating Eq. (8.43) with respect to $\alpha(x)$:

$$\int dy\Gamma_{,\nu}(y)\partial_\nu\delta(x-y) + \int dy\Gamma_i(y)\frac{\delta T_{ij}}{\delta\alpha(x)}\chi_{jo}(y) = 0. \qquad (8.44)$$

Now we define the matrix σ:

$$\sigma\delta(x-y) = \frac{\delta T(x)}{\delta\alpha(y)}\bigg|_{\alpha=0} = \begin{pmatrix} 0 & -1 \\ 1 & 0 \end{pmatrix}\delta(x-y). \qquad (8.45)$$

In this way Eq. (8.44) becomes:

$$\Gamma_i(x)\sigma_{ij}\chi_{jo}(x) - \partial_\nu\Gamma_{,\nu}(x) = 0. \qquad (8.46)$$

Starting from this equation we write the following ones by simply differentiating it with respect to χ_{mo}:

$$\Gamma_{im}(x_1,x_2)\sigma_{ij}\chi_{jo}(x_1) + \Gamma_i(x_1)\sigma_{ij}\delta_{jm}\delta(x_1-x_2) - \partial_\nu^{x_1}\Gamma_{m;\nu}(x_2;x_1) = 0, \quad (8.47)$$

$$\begin{aligned} &\Gamma_{imn}(x_1,x_2,x_3)\sigma_{ij}\chi_{jo}(x_1) \\ &+\Gamma_{im}(x_1,x_2)\sigma_{ij}\delta_{jn}\delta(x_1-x_3) + \Gamma_{in}(x_1,x_3)\sigma_{ij}\delta_{jm}\delta(x_1-x_2) \\ &-\partial_\nu^{x_1}\Gamma_{mn;\nu}(x_2,x_3;x_1) = 0, \end{aligned} \qquad (8.48)$$

and

$$\Gamma_{imnl}(x_1, x_2, x_3, x_4)\sigma_{ij}\chi_{jo}(x_1) + \Gamma_{imn}(x_1, x_2, x_3)\sigma_{ij}\delta_{jl}\delta(x_1 - x_4)$$
$$+\Gamma_{iml}(x_1, x_2, x_4)\sigma_{ij}\delta_{jn}\delta(x_1 - x_3) + \Gamma_{inl}(x_1, x_3, x_4)\sigma_{ij}\delta_{jm}\delta(x_1 - x_2)$$
$$-\partial_\nu^{x_1}\Gamma_{mnl;\nu}(x_2, x_3, x_4; x_1) = 0. \tag{8.49}$$

We can perform now the Fourier transform of these identities:

$$\Gamma(q) \equiv \int dx e^{-iqx}\, \Gamma(x), \tag{8.50}$$

$$\Gamma(x) = \frac{1}{\beta\Omega}\sum_q e^{iqx}\Gamma(q), \tag{8.51}$$

where $qx = \omega\tau + \mathbf{kx}$. We have to deal with a convolution of two differently normalized Fourier transforms:

$$\int dx\, e^{-ikx} f(x)g(x) = (\beta\Omega)^{-1/2}\sum_q f(k - q)g(q) \tag{8.52}$$

and a delta function:

$$\int dx_1 e^{-ik_1 x_1} \int dx_2 e^{-ik_2 x_2} f(x_1)\delta(x_1 - x_2) = f(k_1 + k_2). \tag{8.53}$$

So we obtain for the vertex parts (here $k_\nu = (i\omega, \mathbf{k})$ due to the presence of the i in the ∂_τ term entering (8.29)):

$$(\beta\Omega)^{-1/2}\sum_q \Gamma_i(k - q)\sigma_{ij}\chi_{jo}(q) - ik_\nu\Gamma_{;\nu}(k) = 0. \tag{8.54}$$

$$(\beta\Omega)^{-1/2}\sum_q \Gamma_{im}(k_1 - q, k_2)\sigma_{ij}\chi_{jo}(q) + \Gamma_i(k_1 + k_2)\sigma_{im} - ik_\nu^1\Gamma_{m;\nu}(k_2; k_1) = 0 \tag{8.55}$$

$$(\beta\Omega)^{-1/2}\sum_q \Gamma_{imn}(k_1 - q, k_2, k_3)\sigma_{ij}\chi_{jo}(q)$$
$$+\Gamma_{im}(k_1 + k_3, k_2)\sigma_{in} + \Gamma_{in}(k_1 + k_2, k_3)\sigma_{im}$$
$$-ik_\nu^1\Gamma_{mn;\nu}(k_2, k_3; k_1) = 0 \tag{8.56}$$

$$(\beta\Omega)^{-1/2}\sum_q \Gamma_{imnl}(k_1 - q, k_2, k_3, k_4)\sigma_{ij}\chi_{jo}(q)$$
$$+\Gamma_{imn}(k_1 + k_4, k_2, k_3)\sigma_{il} + \Gamma_{iml}(k_1 + k_3, k_2, k_4)\sigma_{in} + \Gamma_{inl}(k_1 + k_2, k_3, k_4)\sigma_{im}$$
$$-ik_\nu^1\Gamma_{mnl;\nu}(k_2, k_3, k_4; k_1) = 0 \tag{8.57}$$

We can now consider the limit of uniform external sources, which guaranties the invariance under translations. Each vertex part can thus be written as follows:

$$\Gamma(k_1, k_2, \ldots, k_n) = \bar{\Gamma}(k_1, k_2, k_{n-1})\delta_{k_1 + k_2 + \ldots + k_n, 0}\beta\Omega. \tag{8.58}$$

In particular the condensate will be uniform:

$$\chi_{io}(q) = (\beta\Omega)^{1/2}\delta_{q,0}\chi_{io}, \tag{8.59}$$

so we have:

$$\Gamma_i(k)\sigma_{ij}\chi_{jo} - ik_\nu\Gamma_{;\nu}(k) = 0, \tag{8.60}$$

$$\Gamma_{im}(k_1,k_2)\sigma_{ij}\chi_{jo} + \Gamma_i(k_1+k_2)\sigma_{im} - ik_\nu^1\Gamma_{m;\nu}(k_2;k_1) = 0, \tag{8.61}$$

$$\Gamma_{imn}(k_1,k_2,k_3)\sigma_{ij}\chi_{jo} + \Gamma_{im}(k_1+k_3,k_2)\sigma_{in} + \Gamma_{in}(k_1+k_2,k_3)\sigma_{im}$$
$$-ik_\nu^1\Gamma_{mn;\nu}(k_2,k_3;k_1) = 0, \tag{8.62}$$

and

$$\Gamma_{imnl}(k_1,k_2,k_3,k_4)\sigma_{ij}\chi_{jo}$$
$$+\Gamma_{imn}(k_1+k_4,k_2,k_3)\sigma_{il} + \Gamma_{iml}(k_1+k_3,k_2,k_4)\sigma_{in} + \Gamma_{inl}(k_1+k_2,k_3,k_4)\sigma_{im}$$
$$-ik_\nu^1\Gamma_{mnl;\nu}(k_2,k_3,k_4;k_1) = 0. \tag{8.63}$$

Now we rewrite the same identities eliminating the overall delta functions:

$$\bar\Gamma_i(0)\sigma_{ij}\chi_{jo} = 0, \tag{8.64}$$

$$\bar\Gamma_{im}(k_1)\sigma_{ij}\chi_{jo} + \bar\Gamma_i(0)\sigma_{im} - ik_\nu^1\bar\Gamma_{m;\nu}(-k_1) = 0, \tag{8.65}$$

$$\bar\Gamma_{imn}(k_1,k_2)\sigma_{ij}\chi_{jo} + \bar\Gamma_{im}(-k_2)\sigma_{in} + \bar\Gamma_{in}(k_1+k_2)\sigma_{im}$$
$$-ik_\nu^1\bar\Gamma_{mn;\nu}(k_2,-k_1-k_2) = 0, \tag{8.66}$$

and

$$\bar\Gamma_{imnl}(k_1,k_2,k_3)\sigma_{ij}\chi_{jo} + \bar\Gamma_{imn}(-k_2-k_3,k_2)\sigma_{il} + \bar\Gamma_{iml}(k_1+k_3,k_2)\sigma_{in}$$
$$+\bar\Gamma_{inl}(k_1+k_2,k_3)\sigma_{im} - ik_\nu^1\bar\Gamma_{mnl;\nu}(k_2,k_3,-k_1-k_2-k_3) = 0. \tag{8.67}$$

822   Particular Cases of Ward Identities

In this section we write down explicitly various WI that we will need in the future. We consider in particular the case of vanishing external sources and finite χ_{1o}. In this case the Ward identities have the following form:

$$\bar\Gamma_2(0)\chi_{1o} = 0, \tag{8.68}$$

$$\bar\Gamma_{2m}(k_1)\chi_{1o} + \bar\Gamma_i(0)\sigma_{im} - ik_\nu^1\bar\Gamma_{m;\nu}(-k_1) = 0, \tag{8.69}$$

$$\bar\Gamma_{2mn}(k_1,k_2)\chi_{1o} + \bar\Gamma_{im}(-k_2)\sigma_{in} + \bar\Gamma_{in}(k_1+k_2)\sigma_{im}$$
$$-ik_\nu^1\bar\Gamma_{mn;\nu}(k_2,-k_1-k_2) = 0, \tag{8.70}$$

and

$$\bar{\Gamma}_{2mnl}(k_1, k_2, k_3)\chi_{1o} + \bar{\Gamma}_{imn}(-k_2 - k_3, k_2)\sigma_{il} + \bar{\Gamma}_{iml}(k_1 + k_3, k_2)\sigma_{in}$$
$$+\bar{\Gamma}_{inl}(k_1 + k_2, k_3)\sigma_{im} - ik_\nu^1\bar{\Gamma}_{mnl;\nu}(k_2, k_3, -k_1 - k_2 - k_3) = 0. \quad (8.71)$$

We consider in addition particular cases which will be useful below:

$$\bar{\Gamma}_2(0)\chi_{1o} = 0, \quad (8.72)$$

$$
\begin{aligned}
(m = 1) \quad & \bar{\Gamma}_{21}(k_1)\chi_{1o} + \bar{\Gamma}_2(0) - ik_\nu^1\bar{\Gamma}_{1;\nu}(-k_1) &= 0 \\
(m = 2) \quad & \bar{\Gamma}_{22}(k_1)\chi_{1o} - \bar{\Gamma}_1(0) - ik_\nu^1\bar{\Gamma}_{2;\nu}(-k_1) &= 0;
\end{aligned} \quad (8.73)
$$

for $m = 1$ and $n = 1$ we have:

$$\bar{\Gamma}_{211}(k_1, k_2)\chi_{1o} + \bar{\Gamma}_{21}(-k_2) + \bar{\Gamma}_{21}(k_1 + k_2) - ik_\nu^1\bar{\Gamma}_{11;\nu}(k_2, -k_1 - k_2) = 0. \quad (8.74)$$

For $m = 2$ and $n = 1$:

$$\bar{\Gamma}_{221}(k_1, k_2)\chi_{1o} + \bar{\Gamma}_{22}(-k_2) - \bar{\Gamma}_{11}(k_1 + k_2) - ik_\nu^1\bar{\Gamma}_{21;\nu}(k_2, -k_1 - k_2) = 0. \quad (8.75)$$

For $m = 1$, $n = 2$ (the same as above)

$$\bar{\Gamma}_{212}(k_1, k_2)\chi_{1o} - \bar{\Gamma}_{11}(-k_2) + \bar{\Gamma}_{22}(k_1 + k_2) - ik_\nu^1\bar{\Gamma}_{12;\nu}(k_2, -k_1 - k_2) = 0. (8.76)$$

For $m = 2$ and $n = 2$:

$$\bar{\Gamma}_{222}(k_1, k_2)\chi_{1o} - \bar{\Gamma}_{12}(-k_2) - \bar{\Gamma}_{12}(k_1 + k_2) - ik_\nu^1\bar{\Gamma}_{22;\nu}(k_2, -k_1 - k_2) = 0. (8.77)$$

For $m = n = l = 2$:

$$\bar{\Gamma}_{2222}(k_1, k_2, k_3)\chi_{1o} - \bar{\Gamma}_{122}(-k_2 - k_3, k_2) - \bar{\Gamma}_{122}(k_1 + k_3, k_2)$$
$$-\bar{\Gamma}_{122}(k_1 + k_2, k_3) - ik_\nu^1\bar{\Gamma}_{222;\nu}(k_2, k_3, -k_1 - k_2 - k_3) = 0. \quad (8.78)$$

For $m = 2$, $n = l = 1$:

$$\bar{\Gamma}_{2211}(k_1, k_2, k_3)\chi_{1o} + \bar{\Gamma}_{221}(-k_2 - k_3, k_2) + \bar{\Gamma}_{221}(k_1 + k_3, k_2)$$
$$-\bar{\Gamma}_{111}(k_1 + k_2, k_3) - ik_\nu^1\bar{\Gamma}_{211;\nu}(k_2, k_3, -k_1 - k_2 - k_3) = 0. \quad (8.79)$$

For $m = n = 2$ and $l = 1$:

$$\bar{\Gamma}_{2221}(k_1, k_2, k_3)\chi_{1o} + \bar{\Gamma}_{222}(-k_2 - k_3, k_2) - \bar{\Gamma}_{121}(k_1 + k_3, k_2)$$
$$-\bar{\Gamma}_{121}(k_1 + k_2, k_3) - ik_\nu^1\bar{\Gamma}_{221;\nu}(k_2, k_3, -k_1 - k_2 - k_3) = 0 \quad (8.80)$$

823   General Form of Ward Identities

It is possible to obtain a general form of the Ward Identities for all the vertex functions. To this purpose it is convenient to introduce a simplified notations:

$$\Gamma_{n_1,n_2;m} = \bar{\Gamma}_{1...12...2;\mu_1...\mu_m} \,. \tag{8.81}$$

In this way the integers n_1, n_2, and m indicate the number of functional derivatives that have been performed. We can also indicate in a simplified way the local part:

$$-ik_\nu \bar{\Gamma}_{i;\nu} = -k\Gamma_{1;1} .. \tag{8.82}$$

With the help of this notation it is possible to obtain a general expression for the Ward Identities.

We begin by restating in the new notation (8.60):

$$\Gamma_{0,1}\chi_{1o} - \Gamma_{1,0}\chi_{2o} - k\Gamma_{0,0;1} = 0 \,, \tag{8.83}$$

by differentiating $n_2 - 1$ times (8.83) with respect to χ_{2o} we obtain the following expression:

$$\Gamma_{0,n_2}\chi_{1o} - \Gamma_{1,n_2-1}\chi_{2o} - (n_2 - 1)\Gamma_{1,n_2-2} - k\Gamma_{0,n_2-1;1} = 0 \,, \tag{8.84}$$

valid for $n_2 \geq 1$. We can now differentiate with respect to χ_{1o} n_1 times:

$$\Gamma_{n_1,n_2}\chi_{1o} + n_1\Gamma_{n_1-1,n_2} + \Gamma_{n_1+1,n_2-1}\chi_{2o} - (n_2 - 1)\Gamma_{n_1+1,n_2-2} - k\Gamma_{n_1,n_2-1;1} = 0 \,. \tag{8.85}$$

Successive m differentiation by the composite field A_ν lead to the following result:

$$\Gamma_{n_1,n_2;m}\chi_{1o} \quad + \quad n_1\,\Gamma_{n_1-1,n_2;m} + \Gamma_{n_1+1,n_2-1;m}\chi_{2o}$$
$$-(n_2 - 1)\,\Gamma_{n_1+1,n_2-2;m} - k\,\Gamma_{n_1,n_2-1;1+m} = 0 \,. \tag{8.86}$$

The above identities will be useful in the following to give constraints for all the vertex functions of the theory.

Chapter 9

Power Counting and Running Couplings

9.1 Relevance of the Vertex Function

In this Section we change the engineering dimension of the propagators so that it coincide with the IR behavior found at the mean field level. This corresponds to leaving the Bogolubov action dimensionless. In this way we can select the interaction couplings on the basis of their relevance, and the stability of the Bogolubov action can be investigated. We begin by reviewing the simple case of a ϕ^n model, as a guide for the procedure.

911   Engineering Dimensions for ϕ^n

We consider the following Action:

$$S = \int d^d x \left[(\nabla \phi(x))^2 + \sum_n v_n \phi(x)^n \right] \tag{9.1}$$

Measuring all physical quantities in terms of momentum we have $[x] = -1$. As $[S] = 0$ it is easy to find the dimension of the field ϕ and of v_n:

$$-d + 2 + 2[\phi(x)] = 0 \quad \Rightarrow \quad [\phi(x)] = -1 + d/2 \tag{9.2}$$
$$-d + n[\phi(x)] + [v_n] = 0 \quad \Rightarrow \quad [v_n] = d - n(d/2 - 1) \tag{9.3}$$

Now we perform a Fourier transform of the fields:

$$\phi(k) = \frac{1}{V^{1/2}} \int dx\ \phi(x) e^{-ikx} \tag{9.4}$$
$$\phi(x) = \frac{1}{V^{1/2}} \sum_k \phi(k) e^{ikx} \tag{9.5}$$

This changes the dimension of the field:

$$[\phi(k)] = \frac{d}{2} + [\phi(x)] - d = -1. \tag{9.6}$$

In terms of this Fourier transform the Action becomes:

$$S = \sum_k k^2 \phi(k)\phi(-k) + \sum_n \left\{ v_n V^{1-n/2} \sum_{k_1 + \ldots + k_n} \phi(k_1) \ldots \phi(k_n) \right\} \tag{9.7}$$

and the dimension of $w_n \equiv v_n V^{1-n/2}$ is:

$$[w_n] = [v_n] - d(1 - n/2) = n \tag{9.8}$$

We can also find the dimension of the correlation function:

$$\left[\mathcal{G}^{(n)}(x_1, \ldots, x_n)\right] = [\langle \phi(x_1) \ldots \phi(x_n)\rangle] = n\left(\frac{d}{2} - 1\right) \tag{9.9}$$

Now we define the Fourier transform of the correlation function (recall that the definition is *different* from the one used for the fields):

$$\mathcal{G}^{(n)}(k_i) \equiv \int dx_1 \ldots \int dx_n\, \mathcal{G}^{(n)}(x_i) e^{-\sum_j ik_j x_j} \tag{9.10}$$

$$\mathcal{G}^{(n)}(x_i) = V^{-n} \sum_{k_1, \ldots, k_n} \mathcal{G}^{(n)}(k_i) e^{\sum_j ik_j x_j}, \tag{9.11}$$

we always use the properties:

$$\int dx\, e^{ikx} = V\delta_{k,0}, \tag{9.12}$$

$$\sum_k e^{ikx} = V\delta(x). \tag{9.13}$$

With these definitions the dimension of $\mathcal{G}^{(n)}(k_i)$ is the following:

$$[\mathcal{G}^{(n)}(k_i)] = -dn + [\mathcal{G}^{(n)}(x_i)] = -n\left(\frac{d}{2} + 1\right). \tag{9.14}$$

We define now the Green's function without the overall delta function:

$$\mathcal{G}^{(n)}(k_i) = \bar{\mathcal{G}}^{(n)}(k_i)\delta(\sum_i k_i), \tag{9.15}$$

the dimension of a delta function is the inverse of its argument so:

$$[\bar{\mathcal{G}}^{(n)}(k_i)] = [\mathcal{G}^{(n)}(k_i)] + d = -n\left(\frac{d}{2} + 1\right) + d. \tag{9.16}$$

From this expression and the following relation:

$$\bar{\mathcal{G}}^{(n)}(k_i) = \bar{\Gamma}^{(n)}(k_i) \prod_{j=1}^{n} \bar{\mathcal{G}}^{(2)}(k_j) + \dots , \tag{9.17}$$

we can find the dimension of the reduced $(\bar{\Gamma})$ vertex part:

$$[\bar{\Gamma}^{(n)}(k_i)] = [\bar{\mathcal{G}}^{(n)}(k_i)] - n[\bar{\mathcal{G}}^{(2)}(k_i)] = -n(d/2 - 1) + d . \tag{9.18}$$

Note that this dimension coincides with the dimension of the bare interaction coupling v_n in x-space. So we can understand the advantage of this notation: the vertex part $\bar{\Gamma}^{(n)}$ is an intensive quantity like v_n, and often the first term in the perturbative expansion of $\bar{\Gamma}^{(n)}$ is v_n.

9.1.2   Engineering Dimensions for Our Problem

In our case we have some differences with respect to the ϕ^n model presented above. In fact, the integrals are performed over the imaginary time which has the bare dimension of an energy. So we have the following difference: the volume V becomes $\beta\Omega$, where now Ω is the spatial volume and $\beta = 1/k_B T$. The Fourier transforms are defined in (8.14) and (8.15). With arguments similar to the ones used in 9.1.1 we find the dimensions of the various functions:

$$[\psi(x)] = d/2 \tag{9.19}$$
$$[v_n] = -nd/2 + d + 2 \tag{9.20}$$
$$[\mathcal{G}^{(n)}(x_i)] = nd/2 \tag{9.21}$$
$$[\psi(k)] = -1 \tag{9.22}$$
$$[\mathcal{G}^{(n)}(k_i)] = -n(d/2 + 2) \tag{9.23}$$
$$[\bar{\mathcal{G}}^{(n)}(k_i)] = -n(d/2 + 2) + d + 2 \tag{9.24}$$
$$[\bar{\Gamma}^{(n)}(k_i)] = -nd/2 + d + 2 \tag{9.25}$$

These could also be obtained directly from (9.18) for the ϕ^n model by letting $d \to d + 2$.

9.1.3   Elimination of c_o by Rescaling

We now eliminate the dimensional constant $c_o = \sqrt{2\mu}$ from the theory. This will change the dimensions of the fields and of the couplings, but we will have only one dimensional quantity left in the calculation of the vertex functions, *i.e.*, the external momenta if we are studying the infrared (IR) problem, or the cut-off Λ if we are studying the ultraviolet (UV) problem.

We consider the action given in (8.18) and we perform the following rescaling:

$$\begin{cases} \chi_1(\mathbf{k}, \omega) &= \varphi_1(\mathbf{k}, \omega/c_o)/c_o \\ \chi_2(\mathbf{k}, \omega) &= \varphi_2(\mathbf{k}, \omega/c_o) \end{cases}, \qquad (9.26)$$

in this way the leading $k \to 0$ quadratic part of the Action becomes:

$$\begin{aligned} S_Q &= \sum_{\mathbf{k}, \omega} (\tilde{\chi}_1(-\omega, -\mathbf{k}), \tilde{\chi}_2(-\omega, -\mathbf{k})) \begin{pmatrix} -c_o^2 & \omega \\ -\omega & -\mathbf{k}^2 \end{pmatrix} \begin{pmatrix} \tilde{\chi}_1(\omega, \mathbf{k}) \\ \tilde{\chi}_2(\omega, \mathbf{k}) \end{pmatrix} \\ &= \sum_{\mathbf{k}, k_0} (\varphi_1(-k_0, -\mathbf{k}), \varphi_2(-k_0, -\mathbf{k})) \begin{pmatrix} -1 & k_0 \\ -k_0 & -\mathbf{k}^2 \end{pmatrix} \begin{pmatrix} \varphi_1(k_0, \mathbf{k}) \\ \varphi_2(k_0, \mathbf{k}) \end{pmatrix}, \end{aligned}$$
$$(9.27)$$

where $k_0 \equiv \omega/c_o$ This change of variable modifies the dimension of the propagator and of the fields:

$$\bar{\mathcal{G}}_{i,j}^\varphi(\mathbf{k}, k_0) \equiv \langle \varphi_i(-\mathbf{k}, -k_0)\varphi_j(\mathbf{k}, k_0) \rangle = -\frac{1}{2}\frac{1}{\mathbf{k}^2 + k_0^2} \begin{pmatrix} -\mathbf{k}^2 & -k_0 \\ k_0 & -1 \end{pmatrix}. \quad (9.28)$$

In this way:

$$[\bar{\mathcal{G}}_{11}^\varphi(k)] = 0 \qquad [\bar{\mathcal{G}}_{22}^\varphi(k)] = -2 \qquad [\bar{\mathcal{G}}_{12}^\varphi(k)] = -1, \qquad (9.29)$$

and:

$$\begin{cases} [\varphi_1(k)] &= [\tilde{\chi}_1(k)] + 1 &= 0 \\ [\varphi_2(k)] &= [\tilde{\chi}_2(k)] &= -1 \end{cases} \qquad (9.30)$$

We can also write the same transformation for the fields in real space:

$$\begin{cases} \chi_1(\tau, \mathbf{x}) &= \varphi_1(c_o\tau, \mathbf{x})c_o^{-1/2} \\ \chi_2(\tau, \mathbf{x}) &= \varphi_2(c_o\tau, \mathbf{x})c_o^{1/2} \end{cases}, \qquad (9.31)$$

the quadratic action becomes ($x_0 \equiv c_o\tau$ with $[x_0] = -1$):

$$S_Q = \int d\mathbf{x} \int_0^{\beta c} dx_0 \left\{ -\varphi_1(x_0, \mathbf{x})^2 + 2\varphi_1(x_0, \mathbf{x})\partial_{x_0}\varphi_2(x_0, \mathbf{x}) - (\nabla\varphi_2(x_0, \mathbf{x}))^2 \right\}$$
$$(9.32)$$

and the dimension of the fields changes in the following way:

$$\begin{cases} [\varphi_1(x)] &= (d+1)/2 \\ [\varphi_2(x)] &= (d-1)/2 \end{cases} \qquad (9.33)$$

We must also change the definition of the Fourier transform:

$$\varphi_i(k_0, \mathbf{k}) = \frac{1}{(\beta c_o\Omega)^{1/2}} \int_0^{\beta c_o} dx_0 \int d^d\mathbf{x}\varphi_i(x)e^{-ikx} \qquad (9.34)$$

$$\varphi_i(x_0, \mathbf{x}) = \frac{1}{(\beta c_o\Omega)^{1/2}} \sum_{\omega_\nu, \mathbf{k}} \varphi_i(k)e^{ikx}. \qquad (9.35)$$

Note that now the Matsubara frequencies are: $\omega_\nu = 2\pi\nu/(\beta c)$ and the fields are periodic with period βc_o.

With this rescaling we have eliminated c_o from the propagators. In this way the engineering dimension of the propagator are the same of the degree of divergence in the infrared. This fact has important consequences. First of all we can regularize integrals in the UV by using the partial-q trick by 't Hooft and Veltman [97], so we let the UV cut-off to infinity. The integral now contains only the following dimensional quantities: the running couplings and the external momenta. By changing again the variable of integration it is easy to eliminate the internal integrations, so if we scale all the external momenta by a factor α we obtain that the integrations must change by a factor α^{x_d} where x_d is the engineering dimension of the integration. This coincides with the dimension of the diagram only if all the couplings appearing in it are dimensionless. If a non dimensionless coupling appears we can have two different behaviors depending on the dimension of this coupling. If the dimension is negative it is an irrelevant coupling and the degree of divergence of the vertex will be lowered by its presence (*i.e.*, the dimension of the integration must be higher in such a way that the total dimension of the diagram is equal to the dimension of the vertex part to which it contributes). If the dimension of the coupling is positive (relevant coupling) we have stronger divergences in the IR of the diagrams. We consider a simple example. Suppose that $[D(k)] = d - 2$, $[v_1] = 2$, and $[v_2] = -2$. The IR dependence of the following integrals is easily determined by power counting:

$$D(k) = v_1 \int dq \frac{1}{(k-q)^2} \frac{1}{q^2} \sim k^{d-4} \tag{9.36}$$

$$D(k) = \int dq \frac{1}{(k-q)^2} \sim k^{d-2} \tag{9.37}$$

$$D(k) = v_2 \int dq \frac{q^2}{(k-q)^2} \sim k^d \tag{9.38}$$

We can now perform the calculation for the engineering dimensions of the various quantities. We summarize them here:

$$\begin{aligned}
[\varphi_1(x)] &= (d+1)/2 & (9.39)\\
[\varphi_2(x)] &= (d-1)/2 & (9.40)\\
[\varphi_1(k)] &= 0 & (9.41)\\
[\varphi_2(k)] &= -1 & (9.42)\\
\left[\mathcal{G}_{1\ldots12\ldots2}^{(n_1+n_2)}(x_i)\right] &= n_1(d+1)/2 + n_2(d-1)/2 & (9.43)\\
\left[\mathcal{G}_{1\ldots12\ldots2}^{(n_1+n_2)}(k_i)\right] &= -n_1(d+1)/2 - n_2(d+3)/2 & (9.44)\\
\left[\bar{\mathcal{G}}_{1\ldots12\ldots2}^{(n_1+n_2)}(k_i)\right] &= -n_1(d+1)/2 - n_2(d+3)/2 + d + 1 & (9.45)\\
\left[\bar{\Gamma}_{1\ldots12\ldots2}^{(n_1+n_2)}(k_i)\right] &= -n_1(d+1)/2 - n_2(d-1)/2 + d + 1 & (9.46)
\end{aligned}$$

Note that in (9.46) we have restored the old notation for $\bar{\mathcal{G}}^\varphi \to \mathcal{G}$. In (9.46) n_1 stands for the legs of type 1 while n_2 for the legs of type 2. In particular for $d = 3$ we have:

$$\left[\bar{\Gamma}^{(n_1+n_2)}_{1...12...2}(k_i)\right] = 4 - 2n_1 - n_2. \tag{9.47}$$

914   Classication of the Relevant Vertex Functions

We classify now the dimensions of the vertex functions of the theory:

$$\begin{cases} \left[\bar{\Gamma}_1(k_i)\right] &= 2 - \epsilon/2 \\[2mm] \left[\bar{\Gamma}_2(k_i)\right] &= 3 - \epsilon/2 \end{cases} \tag{9.48}$$

$$\begin{cases} \left[\bar{\Gamma}_{11}(k_i)\right] &= 0 \\[2mm] \left[\bar{\Gamma}_{12}(k_i)\right] &= 1 \\[2mm] \left[\bar{\Gamma}_{22}(k_i)\right] &= 2 \end{cases} \quad \begin{cases} \left[\bar{\Gamma}_{111}(k_i)\right] &= -2 + \frac{\epsilon}{2} \\[2mm] \left[\bar{\Gamma}_{112}(k_i)\right] &= -1 + \frac{\epsilon}{2} \\[2mm] \left[\bar{\Gamma}_{122}(k_i)\right] &= \frac{\epsilon}{2} \\[2mm] \left[\bar{\Gamma}_{222}(k_i)\right] &= 1 + \frac{\epsilon}{2} \end{cases} \quad \begin{cases} \left[\bar{\Gamma}_{1111}(k_i)\right] &= -4 + \epsilon \\[2mm] \left[\bar{\Gamma}_{1112}(k_i)\right] &= -3 + \epsilon \\[2mm] \left[\bar{\Gamma}_{1122}(k_i)\right] &= -2 + \epsilon \\[2mm] \left[\bar{\Gamma}_{1222}(k_i)\right] &= -1 + \epsilon \\[2mm] \left[\bar{\Gamma}_{2222}(k_i)\right] &= \epsilon \end{cases} \tag{9.49}$$

Where we have introduced $\epsilon = 3 - d$. In fact, $d = 3$ is the critical dimension at which logarithmic singularities will appear. The relevant vertex functions are thus the following ones (when $\epsilon = 0$):

$$\begin{cases} \bar{\Gamma}_{22}(k_i) & \sim k^2 \\[2mm] \bar{\Gamma}_{12}(k_i), \bar{\Gamma}_{222}(k_i), \dfrac{\partial\bar{\Gamma}_{22}(k_i)}{\partial\omega} & \sim k^1 \\[2mm] \bar{\Gamma}_{11}(k_i), \bar{\Gamma}_{122}(k_i), \bar{\Gamma}_{2222}(k_i), \dfrac{\partial^2\bar{\Gamma}_{22}(k_i)}{\partial\omega^2}, \dfrac{\partial\bar{\Gamma}_{22}(k_i)}{\partial k^2}, \dfrac{\partial\bar{\Gamma}_{222}(k_i)}{\partial\omega} & \sim k^0 \end{cases} \tag{9.50}$$

In terms of ϵ (9.46) becomes:

$$\left[\bar{\Gamma}^{(n_1+n_2)}_{1...12...2}(k_i)\right] = 4 - 2n_1 - n_2 + \epsilon(n_1 + n_2)/2 - \epsilon. \tag{9.51}$$

We do not consider the running coupling which become relevant when we lower the dimension, but restrict the study of the relevant and marginal running coupling at the critical dimension. In fact, note that for $d = 2$ Γ_{tttttt} becomes marginal, and for $d \to 1$ more and more relevant running coupling appears, as shown in Fig. 9.1 We will not consider these running couplings also if we will extend the validity of our approach to $\epsilon < 2$. This procedure

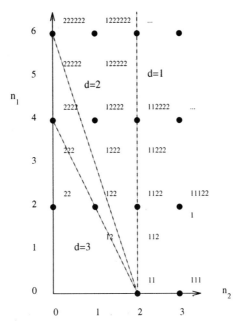

Figure 9.1: Relevant and Irrelevant running coupling by varying the dimension. Points indicate the non vanishing running coupling (even number of type 2 legs) while the dashed lines divide regions of irrelevant ad relevant running couplings for the dimension indicated.

is completely correct in the framework of the epsilon expansion, that is an expansion around the critical dimension, and so need to consider the relevant and marginal running coupling *at* the critical dimension. Of course the validity of the expansion for $\epsilon > 1$ is in general not easy (or possible) to establish. In our case the Ward identities give strong constraints on the whole perturbation theory so that we will be able to give results at *all* order in a epsilon-expansion. In any case we will check explicitly that the the expansion gives the correct result for $d > 1$ within the skeleton equation for the perturbation theory. So we postpone to Section 11.2 the discussion of the additional relevant running coupling (that appear for $d < 2$) after the solution to the problem has been found as an epsilon-expansion around $d = 3$.

9.2   Running Couplings

In view of the RG treatment we write now a general form for the action.

$$-S[\tilde{\chi}_i] =$$
$$= \beta\Omega v_0 + (\beta\Omega)^{1/2} v_1 \tilde{\chi}_1(0) + (\beta\Omega)^{1/2} v_2 \tilde{\chi}_2(0)$$

$$\frac{1}{2!}\sum_k \left\{[v_{11}+z_{11}\mathbf{k}^2]\tilde\chi_1(-k)\tilde\chi_1(k)+2[v_{12}+w_{12}\omega]\tilde\chi_1(-k)\tilde\chi_2(k)\right.$$

$$\left.+[v_{22}+u_{22}\omega^2+z_{22}\mathbf{k}^2]\tilde\chi_2(-k)\tilde\chi_2(k)\right\}$$

$$+\frac{(\beta\Omega)^{-1/2}}{3!}\sum_{k_1+k_2+k_3}\left\{v_{222}\tilde\chi_2(k_1)\tilde\chi_2(k_2)\tilde\chi_2(k_3)+3v_{122}\tilde\chi_1(k_1)\tilde\chi_2(k_2)\tilde\chi_2(k_3)\right.$$

$$\left.+v_{111}\tilde\chi_1(k_1)\tilde\chi_1(k_2)\tilde\chi_1(k_3)\right\}$$

$$+\frac{(\beta\Omega)^{-1}}{4!}\sum_{k_1+k_2+k_3+k_4}\left\{v_{2222}\tilde\chi_2(k_1)\tilde\chi_2(k_2)\tilde\chi_2(k_3)\tilde\chi_2(k_4)\right.$$

$$\left.+6v_{1122}\tilde\chi_1(k_1)\tilde\chi_1(k_2)\tilde\chi_2(k_3)\tilde\chi_2(k_4)+v_{1111}\tilde\chi_1(k_1)\tilde\chi_1(k_2)\tilde\chi_1(k_3)\tilde\chi_1(k_4)\right\}.$$

$$(9.52)$$

In (9.52) we have considered only the relevant or marginal running couplings at $d=3$ and the irrelevant ones present in the bare action. We will soon discard the irrelevant running couplings and discuss in Section 11.2 the effective irrelevance of such terms also for the renormalized perturbation theory. On the other side the form of the action given in (9.52) will be useful in the following for simple checks on the Ward Identities at the lowest order.

Note also that we have not considered the following couplings because their contribution vanishes after the sum over ω_ν:

$$\sum_k w_{22}\omega\tilde\chi_2(-k)\tilde\chi_2(k) = 0, \qquad (9.53)$$

$$(c\beta\Omega)^{1/2}\sum_{k_1+k_2+k_3} w_{222}\omega_1\tilde\chi_2(k_1)\tilde\chi_2(k_2)\tilde\chi_2(k_3) = 0. \qquad (9.54)$$

The origin of this properties lies on the fact that the Fourier transform of this term as a function of x is a total derivative and so can be neglected from the action:

$$\int \chi_2^m(x)\partial_\tau\chi_2(x)d\tau d\mathbf{x} = 0. \qquad (9.55)$$

So at this point the set of retained running couplings is the following:

$$\{v_0,v_1,v_2,v_{11},v_{12},v_{22},w_{12},u_{22},z_{22},z_{11},v_{111},v_{122},v_{222},v_{2222},v_{1122},v_{1111}\}. \qquad (9.56)$$

The quadratic part can be written as follows:

$$S_Q = -\frac{1}{2!}\sum_{ij,k}\tilde\chi_i(-k)Q_{ij}\tilde\chi_j(k), \qquad (9.57)$$

where:

$$Q = \begin{pmatrix} v_{11}+z_{11}\mathbf{k}^2 & , & v_{12}+w_{12}\omega \\ v_{12}-w_{12}\omega & , & v_{22}+u_{22}\omega^2+z_{22}\mathbf{k}^2 \end{pmatrix} \qquad (9.58)$$

By a comparison with Eq. (8.17) we can write the bare values for the running couplings appearing in (9.52). For future reference we report the complete list of bare values of the running coupling:

$$
\begin{aligned}
v_0 &= -(\mu - v\chi_{1o}^2/2)\chi_{1o}^2 &&= -\mu^2/2v \\
v_1 &= -2\chi_{1o}(\mu - v\chi_{1o}^2) &&= 0 \\
v_2 &= 0 &&= 0
\end{aligned}
$$

$$
\begin{aligned}
v_{222} &= 0 &&= 0 \\
v_{122} &= 4v\chi_{1o} &&= 4(\mu v)^{1/2} \\
v_{111} &= 12v\chi_{1o} &&= 12(\mu v)^{1/2}
\end{aligned}
$$

$$
\begin{aligned}
v_{11} &= -2(\mu - 3v\chi_{1o}^2) &&= 4\mu \\
v_{12} &= 0 &&= 0 \\
v_{21} &= 0 &&= 0 \\
v_{22} &= -2(\mu - v\chi_{1o}^2) &&= 0 \\
w_{12} &= -2 &&= -2 \\
u_{22} &= 0 &&= 0 \\
z_{22} &= 2 &&= 2 \\
z_{11} &= 2 &&= 2
\end{aligned}
$$

$$
\begin{aligned}
v_{2222} &= 12v &&= 12v \\
v_{1122} &= 4v &&= 4v \\
v_{1111} &= 12v &&= 12v
\end{aligned}
$$

$$(9.59)$$

The second column of (9.59) refers to the bare value, the third refers to the bare value at the mean-field level, *i.e.*, with $\chi_{1o}^2 = \mu/v$.

921   Symmetry of $\mathcal{G}$ and Γ

In this Section we consider the constraints given by Time Reversal, Parity, Space Rotations, and Exchange of fields components. In particular we begin with a discussion of the symmetry properties of the single-particle propagator. For this purpose we write in detail equation (8.40)

$$\int dz \Gamma_{ii'}(x, z)\mathcal{G}_{i'j}(z, y) = \delta_{ij}\delta(x - y), \quad (9.60)$$

and its Fourier transform with the help of (8.50) and (8.51):

$$
\begin{aligned}
\int dx \int dy \, e^{-ik_1 x - ik_2 y}\delta_{ij}\delta(x - y) &= \beta\Omega\delta_{ij}\delta_{k_1+k_2,0} \\
&= \int dz \Gamma_{ii'}(k_1, z)\mathcal{G}_{i'j}(z, k_2) \\
&= \int dz (\beta\Omega)^{-2} \sum_{qq'} \Gamma_{ii'}(k_1, q)\mathcal{G}_{i'j}(q', k_2)e^{i(q+q')z} \\
&= (\beta\Omega)^{-1} \sum_{q} \Gamma_{ii'}(k_1, q)\mathcal{G}_{i'j}(-q, k_2) \\
&= \beta\Omega\bar{\Gamma}_{ii'}(k_1)\bar{\mathcal{G}}_{i'j}(k_1)\delta_{k_1+k_2,0} .
\end{aligned}
\quad (9.61)
$$

Eliminating the delta functions we obtain:

$$\bar{\Gamma}_{ij}(k) = \left[\bar{\mathcal{G}}^{-1}(k)\right]_{ij} \cdot \cdot , \quad (9.62)$$

so we can write the explicit form of (9.62):

$$\begin{pmatrix} \Gamma_{11}(k) & \Gamma_{12}(k) \\ \Gamma_{21}(k) & \Gamma_{22}(k) \end{pmatrix} = \begin{pmatrix} \mathcal{G}_{22}(k) & -\mathcal{G}_{12}(k) \\ -\mathcal{G}_{21}(k) & \mathcal{G}_{11}(k) \end{pmatrix} \frac{1}{Z(k)}, \tag{9.63}$$

where

$$Z(k) = \mathcal{G}_{11}(k)\mathcal{G}_{22}(k) - \mathcal{G}_{12}(k)\mathcal{G}_{21}(k). \tag{9.64}$$

We now examine the symmetry properties of $\mathcal{G}$:

$$\mathcal{G}_{ii}(x_1, x_2) = \langle \chi_i(x_1)\chi_i(x_2) \rangle_C = \mathcal{G}_{ii}(x_2, x_1), \tag{9.65}$$

by Fourier transforming we obtain:

$$\mathcal{G}_{ii}(k_1, k_2) = \mathcal{G}_{ii}(k_2, k_1) \quad \Rightarrow \quad \mathcal{G}_{ii}(k) = \mathcal{G}_{ii}(-k). \tag{9.66}$$

In general we can write:

$$\mathcal{G}_{ij}(k, -k) = \langle \chi_i(k)\chi_j(-k) \rangle_C = \langle \chi_j(-k)\chi_i(k) \rangle_C = \mathcal{G}_{ji}(-k, k) \tag{9.67}$$

So that we obtain the general relation:

$$\mathcal{G}_{ij}(k) = \mathcal{G}_{ji}(-k). \tag{9.68}$$

From Eq. (9.68) and (9.64) we can see easily that the $Z(k)$ is an even function of its argument:

$$Z(-k) = Z(k), \tag{9.69}$$

in this way we obtain that $\Gamma_{ij}(k)$ satisfy the same symmetry properties of $\mathcal{G}_{ij}(k)$:

$$\Gamma_{ij}(k) = \Gamma_{ji}(-k). \tag{9.70}$$

Parity (P)

The system is invariant under parity. In the functional integral formalism this means that the action S is invariant under the following transformation:

$$\begin{cases} \chi_i(\tau, \mathbf{x}) \rightarrow \chi_i(\tau, -\mathbf{x}) \\ \mathbf{A}(\tau, \mathbf{x}) \rightarrow -\mathbf{A}(\tau, -\mathbf{x}), \end{cases} \tag{9.71}$$

or more explicitly:

$$S[\chi_i(\tau, \mathbf{x}), \mathbf{A}(\tau, \mathbf{x})] = S[\chi_i(\tau, -\mathbf{x}), -\mathbf{A}(\tau, -\mathbf{x})] \tag{9.72}$$

and for the connected Green's functions we obtain:

$$\begin{aligned} \mathcal{G}_{ij}(\tau_1, \mathbf{x}_1; \tau_2, \mathbf{x}_2) &= \langle \chi_i(\tau_1, \mathbf{x}_1)\chi_j(\tau_2, \mathbf{x}_2) \rangle \\ &= \langle \chi_i(\tau_1, -\mathbf{x}_1)\chi_j(\tau_2, -\mathbf{x}_2) \rangle \\ &= \mathcal{G}_{ij}(\tau_1, -\mathbf{x}_1; \tau_2, -\mathbf{x}_2). \end{aligned} \tag{9.73}$$

So in momentum space we have:

$$\mathcal{G}_{ij}(\omega, \mathbf{k}) \overset{P}{=} \mathcal{G}_{ij}(\omega, -\mathbf{k}) \tag{9.74}$$

$$\mathcal{G}_{i;l}(\omega, \mathbf{k}) \overset{P}{=} -\mathcal{G}_{i;l}(\omega, -\mathbf{k}) \tag{9.75}$$

$$\mathcal{G}_{ij;l}(\{\omega\}, \{\mathbf{k}\}) \overset{P}{=} -\mathcal{G}_{ij;l}(\{\omega\}, \{-\mathbf{k}\}) . \tag{9.76}$$

From (9.71) it is easy to find similar relations for the other Green's functions.

Time Reversal (T)

Time reversal symmetry means that the action (8.9) is invariant under the following transformation:

$$\begin{cases} \chi_1(\tau, \mathbf{x}) & \to & \chi_1(-\tau, \mathbf{x}) \\ \chi_2(\tau, \mathbf{x}) & \to & -\chi_2(-\tau, \mathbf{x}) \\ \mathbf{A}(\tau, \mathbf{x}) & \to & -\mathbf{A}(-\tau, \mathbf{x}) \\ \mu(\tau, \mathbf{x}) & \to & \mu(-\tau, \mathbf{x}) \\ \lambda_1(\tau, \mathbf{x}) & \to & \lambda_1(-\tau, \mathbf{x}) \\ \lambda_2(\tau, \mathbf{x}) & \to & -\lambda_2(-\tau, \mathbf{x}) . \end{cases} \tag{9.77}$$

The change of sign of $\mathbf{A}$ corresponds to the change of sign of the charges' velocities that generate the magnetic fields. This invariance gives the following equation for $\mathcal{G}$:

$$\mathcal{G}_1(x) = \langle \chi_1(x) \rangle_{\lambda_1=0^+, \lambda_2=0} \overset{TP}{=} \langle \chi_1(-x) \rangle_{\lambda_1=0^+, \lambda_2=0} \tag{9.78}$$

$$\mathcal{G}_2(x) = \langle \chi_2(x) \rangle_{\lambda_1=0^+, \lambda_2=0} \overset{TP}{=} -\langle \chi_2(-x) \rangle_{\lambda_1=0^+, \lambda_2=0} , \tag{9.79}$$

so that the Fourier transform becomes:

$$\mathcal{G}_1(k) = \mathcal{G}_1(-k) \overset{k\to 0}{=} \mathcal{G}_1(0) = \chi_{1o} \tag{9.80}$$

$$\mathcal{G}_2(k) = -\mathcal{G}_2(-k) \overset{k\to 0}{=} \mathcal{G}_2(0) = 0 . \tag{9.81}$$

Note that this result depends strongly on the fact that $\lambda_2 \to 0$ before λ_1.

We can obtain useful properties also for the other $\mathcal{G}$'s:

$$\begin{cases} \mathcal{G}_{11}(\omega, \mathbf{k}) & \overset{T}{=} & \mathcal{G}_{11}(-\omega, \mathbf{k}) \\ \mathcal{G}_{12}(\omega, \mathbf{k}) & \overset{T}{=} & -\mathcal{G}_{12}(-\omega, \mathbf{k}) \\ \mathcal{G}_{111}(k_1, k_2) & \overset{PT}{=} & \mathcal{G}_{111}(-k_1, -k_2) \\ \mathcal{G}_{122}(k_1, k_2) & \overset{PT}{=} & \mathcal{G}_{122}(-k_1, -k_2) \\ \mathcal{G}_{222}(k_1, k_2) & \overset{PT}{=} & -\mathcal{G}_{222}(-k_1, -k_2) \\ \mathcal{G}_{2222}(k_1, k_2, k_3) & \overset{PT}{=} & \mathcal{G}_{2222}(-k_1, -k_2, -k_3) , \end{cases} \tag{9.82}$$

where we used also the invariance under Parity. We can obtain similar properties for the composite Green's function by considering also the inversion of sign of $\mathbf{A}$ and the symmetry properties of the composite part:

$$S_{;0}(\tau, \mathbf{x}) \overset{T}{\to} S_{;0}(-\tau, \mathbf{x}) \tag{9.83}$$
$$S_{;0}(\tau, \mathbf{x}) \overset{P}{\to} S_{;0}(\tau, -\mathbf{x}) \tag{9.84}$$
$$S_{;l}(\tau, \mathbf{x}) \overset{T}{\to} -S_{;l}(-\tau, \mathbf{x}) \tag{9.85}$$
$$S_{;l}(\tau, \mathbf{x}) \overset{P}{\to} -S_{;l}(\tau, -\mathbf{x}). \tag{9.86}$$

Note also the following subtle point:

$$\mathcal{G}_{1;l}(\omega, \mathbf{k})|_{\mathbf{A}} \overset{T}{=} - \mathcal{G}_{1;l}(-\omega, \mathbf{k})|_{-\mathbf{A}} \overset{P}{=} \mathcal{G}_{1;l}(-\omega, -\mathbf{k})|_{\mathbf{A}} \tag{9.87}$$

So we have for vanishing external Sources:

$$\begin{cases} \mathcal{G}_{1;0}(\omega, \mathbf{k}) \overset{T}{=} \mathcal{G}_{1;0}(-\omega, \mathbf{k}) \\ \mathcal{G}_{2;0}(\omega, \mathbf{k}) \overset{T}{=} -\mathcal{G}_{2;0}(-\omega, \mathbf{k}) \\ \mathcal{G}_{1;l}(\omega, \mathbf{k}) \overset{T}{=} -\mathcal{G}_{1;l}(-\omega, \mathbf{k}) \\ \mathcal{G}_{2;l}(\omega, \mathbf{k}) \overset{T}{=} \mathcal{G}_{2;l}(-\omega, \mathbf{k}) \end{cases} \tag{9.88}$$

and:

$$\begin{cases} \mathcal{G}_{;00}(k) = \mathcal{G}_{;00}(-k) \\ \mathcal{G}_{;0l}(k) = -\mathcal{G}_{;0l}(-\omega, \mathbf{k}) = \mathcal{G}_{;0l}(-k) \\ \mathcal{G}_{;lm}(k) = \mathcal{G}_{;lm}(-\omega, \mathbf{k}) = \mathcal{G}_{;lm}(-k) \end{cases} \tag{9.89}$$

$$\begin{cases} \mathcal{G}_{11;0}(k_1, k_2) \overset{T}{=} \mathcal{G}_{11;0}(-\omega_1, \mathbf{k}_1, -\omega_2, \mathbf{k}_2) \overset{P}{=} \mathcal{G}_{11;0}(-\omega_1, -\mathbf{k}_1, -\omega_2, -\mathbf{k}_2) \\ \mathcal{G}_{22;0}(k_1, k_2) \overset{T}{=} \mathcal{G}_{22;0}(-\omega_1, \mathbf{k}_1, -\omega_2, \mathbf{k}_2) \overset{P}{=} \mathcal{G}_{22;0}(-\omega_1, -\mathbf{k}_1, -\omega_2, -\mathbf{k}_2) \\ \mathcal{G}_{12;0}(k_1, k_2) \overset{T}{=} -\mathcal{G}_{12;0}(-\omega_1, \mathbf{k}_1, -\omega_2, \mathbf{k}_2) \overset{P}{=} -\mathcal{G}_{12;0}(-\omega_1, -\mathbf{k}_1, -\omega_2, -\mathbf{k}_2) \end{cases} \tag{9.90}$$

$$\begin{cases} \mathcal{G}_{11;l}(k_1, k_2) \overset{T}{=} -\mathcal{G}_{11;l}(-\omega_1, \mathbf{k}_1, -\omega_2, \mathbf{k}_2) \overset{P}{=} \mathcal{G}_{11;l}(-\omega_1, -\mathbf{k}_1, -\omega_2, -\mathbf{k}_2) \\ \mathcal{G}_{22;l}(k_1, k_2) \overset{T}{=} -\mathcal{G}_{22;l}(-\omega_1, \mathbf{k}_1, -\omega_2, \mathbf{k}_2) \overset{P}{=} \mathcal{G}_{22;l}(-\omega_1, -\mathbf{k}_1, -\omega_2, -\mathbf{k}_2) \\ \mathcal{G}_{12;l}(k_1, k_2) \overset{T}{=} \mathcal{G}_{12;l}(-\omega_1, \mathbf{k}_1, -\omega_2, \mathbf{k}_2) \overset{P}{=} -\mathcal{G}_{12;l}(-\omega_1, -\mathbf{k}_1, -\omega_2, -\mathbf{k}_2) \end{cases} \tag{9.91}$$

Exchange of the Two Components (X)

We note that the action (8.9) is invariant also under the following transformation (X):

$$\begin{cases} \chi_1(x) \to \chi_2(-x) \\ \chi_2(x) \to \chi_1(-x) \\ \lambda_1(x) \to \lambda_2(-x) \\ \lambda_2(x) \to \lambda_1(-x). \end{cases} \tag{9.92}$$

This invariance gives the following conditions on the $\mathcal{G}$:

$$
\begin{aligned}
\mathcal{G}_{ij;l}(k_1, k_2) &= \mathcal{G}_{ji;l}(k_2, k_1) \\
&\overset{X}{=} \mathcal{G}_{ij;l}(-k_2, -k_1) \\
&\overset{T}{=} -(-1)^{i+j}\mathcal{G}_{ij;l}(\omega_2, -\mathbf{k}_2, \omega_1, -\mathbf{k}_1) \\
&\overset{P}{=} (-1)^{i+j}\mathcal{G}_{ij;l}(\omega_2, \mathbf{k}_2, \omega_1, \mathbf{k}_1) .
\end{aligned}
\tag{9.93}
$$

So we have:

$$
\begin{aligned}
\mathcal{G}_{11;l}(\omega_1, \mathbf{k}_1, \omega_2, \mathbf{k}_2) &= \mathcal{G}_{11;l}(\omega_2, \mathbf{k}_2, \omega_1, \mathbf{k}_1) \\
\mathcal{G}_{12;l}(\omega_1, \mathbf{k}_1, \omega_2, \mathbf{k}_2) &= -\mathcal{G}_{12;l}(\omega_2, \mathbf{k}_2, \omega_1, \mathbf{k}_1) ,
\end{aligned}
$$

$$\tag{9.94}$$
$$\tag{9.95}$$

and

$$
\begin{aligned}
\mathcal{G}_{11;0}(k_1, k_2) &= \mathcal{G}_{11;0}(k_2, k_1) \\
\mathcal{G}_{12;0}(k_1, k_2) &= \mathcal{G}_{21;0}(k_2, k_1) \\
&\overset{X}{=} \mathcal{G}_{12;0}(-k_2, -k_1) \\
&\overset{T}{=} -\mathcal{G}_{12;0}(\omega_2, -\mathbf{k}_2, \omega_1, -\mathbf{k}_1) \\
&\overset{P}{=} -\mathcal{G}_{12;0}(k_2, k_1) .
\end{aligned}
\tag{9.96}
$$

$$\tag{9.97}$$

Conditions on the Form of the Green's Functions Given by the Symmetry

From above results we can write the Green's functions (or the vertex parts) in such a way that their symmetry properties are explicitly shown.

We consider, for example, a function that is completely symmetric under a change of sign of $\mathbf{k}$ and separately of ω:

$$
f(\omega, \mathbf{k}) = f(-\omega, \mathbf{k}) = f(\omega, -\mathbf{k}).
\tag{9.98}
$$

We ask whether it can be written also in the following way:

$$
f(\omega, \mathbf{k}) = f(\omega^2, \mathbf{k}^2).
\tag{9.99}
$$

We expect that for the spatial part we should have no problem due also to the rotational symmetry of the system. For the frequency part one may ask whether a branch cut is present in the complex plane, and the resulting function becomes:

$$
f(\omega, \mathbf{k}) = f(|\omega|, \mathbf{k}^2).
\tag{9.100}
$$

To understand this point we can have a look to a typical symmetric diagram:

$$
f(\omega, \mathbf{k}) = \int dq_0 \int d^d\mathbf{q} \frac{1}{(q_0^2 + \mathbf{q}^2)[(q_0 - \omega)^2 + (\mathbf{q} - \mathbf{k})^2]} .
\tag{9.101}
$$

We note that for given $\mathbf{k}$ finite the branch cut begins at $|\omega| > \mathbf{k}^2$ so that for $\omega = 0$ we have no cut. When $\mathbf{k} = 0$ there is actually a cut, but only for $\mathbf{q} = 0$, and $\mathbf{q}$ is integrated in $d > 1$ dimensions, so the contribution of the cut to the $f(\omega, \mathbf{k})$ is zero. For these reasons we expect that we can always write the symmetric functions as functions only of the variables squared. Of course this fact does not prevent the Green's functions to be singular in their dependence on ω^2, as they effectively are.

So we can write each Green's function in the following way:

$$\begin{cases} \mathcal{G}_1(k) &= \mathcal{G}_1(0) = \chi_{1o} \\ \mathcal{G}_2(k) &= \mathcal{G}_2(0) = 0 \\ \mathcal{G}_{11}(k) &= \mathcal{F}_{11}(\omega^2, \mathbf{k}^2) \\ \mathcal{G}_{22}(k) &= \mathcal{F}_{22}(\omega^2, \mathbf{k}^2) \\ \mathcal{G}_{12}(k) &= \omega \mathcal{F}_{12}(\omega^2, \mathbf{k}^2), \end{cases} \tag{9.102}$$

$$\begin{cases} \mathcal{G}_{1;0}(k) &= \mathcal{F}_{1;0}(\omega^2, \mathbf{k}^2) \\ \mathcal{G}_{2;0}(k) &= \omega \mathcal{F}_{2;0}(\omega^2, \mathbf{k}^2) \\ \mathcal{G}_{1;l}(k) &= (i\mathbf{k}_l)\omega \mathcal{F}_{1;v}(\omega^2, \mathbf{k}^2) \\ \mathcal{G}_{2;l}(k) &= (i\mathbf{k}_l)\mathcal{F}_{2;v}(\omega^2, \mathbf{k}^2), \end{cases} \tag{9.103}$$

where by assumption $\mathcal{F}$ is a symmetric function of its arguments $\mathcal{F}(k) = \mathcal{F}(\omega^2, \mathbf{k}^2)$. In the case of more complex momentum dependence we define always $\mathcal{F}$ as as a function of only the even combinations of the variables, for example:

$$\mathcal{F}(k_1, k_2) = \mathcal{F}(\omega_1^2, \omega_2^2, \omega_1\omega_2, \mathbf{k}_1^2, \mathbf{k}_2^2, \mathbf{k}_1\mathbf{k}_2) \tag{9.104}$$

In this way we can write:

$$\begin{cases} \mathcal{G}_{11;0}(k_1, k_2) &= \mathcal{F}_{11;0}(k_1, k_2) \\ \mathcal{G}_{22;0}(k_1, k_2) &= \mathcal{F}_{22;0}(k_1, k_2) \\ \mathcal{G}_{12;0}(k_1, k_2) &= (\omega_1 - \omega_2)\mathcal{F}_{12;0}(k_1, k_2) \\ \mathcal{G}_{11;l}(k_1, k_2) &= (\omega_1 + \omega_2)i(\mathbf{k}_1 + \mathbf{k}_2)_l \mathcal{F}_{11;v}(k_1, k_2) \\ \mathcal{G}_{22;l}(k_1, k_2) &= (\omega_1 + \omega_2)i(\mathbf{k}_1 + \mathbf{k}_2)_l \mathcal{F}_{22;v}(k_1, k_2) \\ \mathcal{G}_{12;l}(k_1, k_2) &= i(\mathbf{k}_1 - \mathbf{k}_2)_l \mathcal{F}_{12;v}(k_1, k_2). \end{cases} \tag{9.105}$$

Exactly the same properties hold for the corresponding vertex parts, and thus for the corresponding insertions. In this case we indicate the symmetric functions with $\mathcal{P}$:

$$\begin{cases} \bar{\Gamma}_1(k) &= \bar{\Gamma}_1(0) = 0^+ \\ \bar{\Gamma}_2(k) &= \bar{\Gamma}_2(0) = 0 \\ \bar{\Gamma}_{11}(k) &= \mathcal{P}_{11}(\omega^2, \mathbf{k}^2) \\ \bar{\Gamma}_{22}(k) &= \mathcal{P}_{22}(\omega^2, \mathbf{k}^2) \\ \bar{\Gamma}_{12}(k) &= \omega \mathcal{P}_{12}(\omega^2, \mathbf{k}^2) \end{cases} \tag{9.106}$$

$$\begin{cases} \bar{\Gamma}_{1;0}(k) &= \mathcal{P}_{1;0}(\omega^2, \mathbf{k}^2) \\ \bar{\Gamma}_{2;0}(k) &= \omega \mathcal{P}_{2;0}(\omega^2, \mathbf{k}^2) \\ \bar{\Gamma}_{1;l}(k) &= (i\mathbf{k}_l)\omega \mathcal{P}_{1;v}(\omega^2, \mathbf{k}^2) \\ \bar{\Gamma}_{2;l}(k) &= (i\mathbf{k}_l)\mathcal{P}_{2;v}(\omega^2, \mathbf{k}^2) \end{cases} \tag{9.107}$$

$$
\begin{cases}
\bar{\Gamma}_{11;0}(k_1, k_2) &= \mathcal{P}_{11;0}(k_1, k_2) \\
\bar{\Gamma}_{22;0}(k_1, k_2) &= \mathcal{P}_{22;0}(k_1, k_2) \\
\bar{\Gamma}_{12;0}(k_1, k_2) &= (\omega_1 - \omega_2)\mathcal{P}_{12;0}(k_1, k_2) \\
\bar{\Gamma}_{11;l}(k_1, k_2) &= (\omega_1 + \omega_2)i(\mathbf{k}_1 + \mathbf{k}_2)_l \mathcal{P}_{11;v}(k_1, k_2) \\
\bar{\Gamma}_{22;l}(k_1, k_2) &= (\omega_1 + \omega_2)i(\mathbf{k}_1 + \mathbf{k}_2)_l \mathcal{P}_{22;v}(k_1, k_2) \\
\bar{\Gamma}_{12;l}(k_1, k_2) &= i(\mathbf{k}_1 - \mathbf{k}_2)_l \mathcal{P}_{12;v}(k_1, k_2) .
\end{cases}
\tag{9.108}
$$

Dimensions of the Symmetric ($\mathcal{P}$) Functions

It is clear that the IR dimension of the symmetric functions is very important, so we report for convenience the following table:

$$
\begin{cases}
[\mathcal{P}_{11}] &= 0 \\
[\mathcal{P}_{12}] &= 0 \\
[\mathcal{P}_{22}] &= 2
\end{cases}
\begin{cases}
[\mathcal{P}_{1;0}] &= -\epsilon/2 \\
[\mathcal{P}_{2;0}] &= -\epsilon/2 \\
[\mathcal{P}_{1;v}] &= -\epsilon/2 \\
[\mathcal{P}_{2;v}] &= 2 - \epsilon/2
\end{cases}
\begin{cases}
[\mathcal{P}_{11;0}] &= -2 \\
[\mathcal{P}_{12;0}] &= -2 \\
[\mathcal{P}_{22;0}] &= 0 \\
[\mathcal{P}_{11;v}] &= -2 \\
[\mathcal{P}_{12;v}] &= 0 \\
[\mathcal{P}_{22;v}] &= 0
\end{cases}
\tag{9.109}
$$

922   Propagator Vertex Part and Fluctuation Matrix

We need a few words to clarify the relationship between the free propagator, the quadratic vertex part, and the fluctuation matrix Q_{ij} when the fields are not independent. It can be shown in general that the propagator has the following expression:

$$
\mathcal{G}(k) = \left[\frac{Q(k) + Q^t(-k)}{2} \right]^{-1},
\tag{9.110}
$$

so in our case where $Q_{ij}(k) = Q_{ji}(-k)$ we have simply:

$$
\mathcal{G}_{ij}(k) = \left[Q(k)^{-1} \right]_{ij}.
\tag{9.111}
$$

The relationship between $\Gamma_{ij}(k)$ and $\mathcal{G}_{ij}(k)$ found before gives the following equation:

$$
\Gamma_{ij}(k) = Q_{ij}(k).
\tag{9.112}
$$

Note that we introduced the factor -1/2 in the definition (9.57) of Q to have this relation as simple as possible.

923   Operative Denition of the Vertex Parts

So far we have defined the vertex parts as the functional derivatives of the Legendre transformation of the Free energy (see Sec. 8.2). We must remember that our perturbation theory is useful only to calculate the Green's functions, i.e., the functional derivatives of F and not of Γ. Luckily we have a simple operative way to calculate the vertex parts, we can in fact demonstrate that they are the 1-particle irreducible part of the corresponding Green's function. We review the demonstration because for the composite operators, due to the presence of the condensate, we have less standard than usual relations.

We begin with the definition of $\Gamma[\chi_o]$ in a simplified notation (here $i \equiv (i, x)$):

$$\Gamma[\chi_o, A] = \chi_{io}\lambda_i[\chi_o, A] - F[\lambda[\chi_o, A], A], \tag{9.113}$$

where the equation that links χ_{1o} to λ is:

$$\chi_{io} = \left.\frac{\delta F}{\delta \lambda_i}\right|_{\lambda, A} \equiv \mathcal{G}_i. \tag{9.114}$$

From Eq. (9.113) and (9.114) we can demonstrate:

$$\Gamma_i \equiv \left.\frac{\delta \Gamma}{\delta \chi_{io}}\right|_{\chi_o, A} = \lambda_i \tag{9.115}$$

$$\Gamma_{;\mu} \equiv \left.\frac{\delta \Gamma}{\delta A_\mu}\right|_{\chi_o, A} = -\left.\frac{\delta F}{\delta A_\mu}\right|_{\lambda, A} \equiv -\mathcal{G}_{;\mu}. \tag{9.116}$$

Non Composite Operators

Eq. (9.115) and Eq. (9.116) give the first operative definitions of the vertex parts. By differentiating (9.115) with respect to λ_l we obtain:

$$\Gamma_{ij}\mathcal{G}_{jl} = \delta_{il} \quad \Rightarrow \quad \Gamma_{ij} = \left(\mathcal{G}^{-1}\right)_{ij}. \tag{9.117}$$

Now this equation can be written in terms of a perturbative expansion where the free action is given by the quadratic part (i.e. by $\mathcal{G}^o$):

$$\mathcal{G} = \mathcal{G}^o + \mathcal{G}^o\Sigma\mathcal{G}^o + \mathcal{G}^o\Sigma\mathcal{G}^o\Sigma\mathcal{G}^o + \ldots = \mathcal{G}^o\left(1 + \Sigma\mathcal{G}\right), \tag{9.118}$$

where Σ is the 1-particle irreducible set of diagrams. By solving (9.118) we obtain the Dyson equation:

$$\mathcal{G}^{-1} = (\mathcal{G}^o)^{-1} - \Sigma, \tag{9.119}$$

so Eq. (9.117) and Eq. (9.119) give:

$$\Gamma_{ij} = (\mathcal{G}^o)^{-1}_{ij} - \Sigma_{ij}. \tag{9.120}$$

Eq. (9.120) constitutes the operative definition of Γ_{ij}.

We differentiate now the first equation in Eq. (9.117) with respect to χ_{ok}:

$$\Gamma_{ijk}\mathcal{G}_{jl} + \Gamma_{ij}\mathcal{G}_{jlm}\Gamma_{mk} = 0 \quad \Rightarrow \quad \Gamma_{ijk} = -\Gamma_{ii'}\Gamma_{jj'}\Gamma_{kk'}\mathcal{G}_{i'j'k'} \tag{9.121}$$

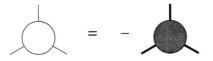

Figure 9.2: Diagrammatic relation between $\mathcal{G}$ and Γ for the 3- and 4-point Green's functions. The heavy lines represent exact single particle Green's functions, while the light lines are simply to indicate the insertion of a particle line.

or also:

$$\mathcal{G}_{ijk} = -\mathcal{G}_{ii'}\mathcal{G}_{jj'}\mathcal{G}_{kk'}\Gamma_{i'j'k'} \qquad (9.122)$$

So Eq. (9.121) assures that Γ_{ijk} is *minus* the 1-particle irreducible part of $\mathcal{G}_{ijk}$. We can write similar equations for the other vertex part and we can see that this is a general statement.

We show also the 4-point $\mathcal{G}$:

$$\begin{aligned}
\mathcal{G}_{ijkl} = & -\mathcal{G}_{ii'}\mathcal{G}_{jj'}\mathcal{G}_{kk'}\mathcal{G}_{ll'}\Gamma_{i'j'k'l'} \\
& -\mathcal{G}_{ii'l}\mathcal{G}_{jj'}\mathcal{G}_{kk'}\Gamma_{i'j'k'} - \mathcal{G}_{ii'}\mathcal{G}_{jj'l}\mathcal{G}_{kk'}\Gamma_{i'j'k'} - \mathcal{G}_{ii'}\mathcal{G}_{jj'}\mathcal{G}_{kk'l}\Gamma_{i'j'k'} \quad (9.123)
\end{aligned}$$

The 3-point Γ in Eq. (9.123) eliminate the internal 1-particle *reducible* diagrams in $\mathcal{G}_{ijkl}$ while the last term is the definition of the 1-particle irreducible Γ function. In this way we can demonstrate by induction that the Vertex parts are 1-particle irreducible.

We report for convenience also the following expression for $\mathcal{G}_{ijkl}$ in terms only of vertex functions and single-particle propagators:

$$\begin{aligned}
\mathcal{G}_{ijkl} = & -\mathcal{G}_{ii'}\mathcal{G}_{jj'}\mathcal{G}_{kk'}\mathcal{G}_{ll'}\Gamma_{i'j'k'l'} + \mathcal{G}_{ii_1}\mathcal{G}_{i'i_2}\mathcal{G}_{ll'}\Gamma_{i_1,i_2,l'}\mathcal{G}_{jj'}\mathcal{G}_{kk'}\Gamma_{i'j'k'} \\
& +\mathcal{G}_{ii'}\mathcal{G}_{jj_1}\mathcal{G}_{j'j_2}\mathcal{G}_{ll'}\Gamma_{j_1j_2l'}\mathcal{G}_{kk'}\Gamma_{i'j'k'} + \mathcal{G}_{ii'}\mathcal{G}_{jj'}\mathcal{G}_{kk_1}\mathcal{G}_{k'k_2'}\mathcal{G}_{ll'}\Gamma_{k_1k_2'l'}\Gamma_{i'j'k'}
\end{aligned}$$

$$(9.124)$$

Composite Operators

We consider now the composite vertex functions. We differentiate Eq. (9.115) with respect to A_μ at χ_o constant, to obtain:

$$\left.\frac{\delta\Gamma_i}{\delta A_\mu}\right|_{\chi_o} \equiv \Gamma_{i;\mu} = \left.\frac{\delta\lambda_i}{\delta A_\mu}\right|_{\chi_o}, \tag{9.125}$$

and from (9.114):

$$\left.\frac{\delta\mathcal{G}_i}{\delta A_\mu}\right|_\lambda \equiv \mathcal{G}_{i;\mu} = \left.\frac{\delta\chi_{io}}{\delta A_\mu}\right|_\lambda. \tag{9.126}$$

Now we perform the derivative with respect to A_μ but with the conjugate variable λ fixed:

$$0 = \left.\frac{\delta\Gamma_i}{\delta A_\mu}\right|_\lambda = \left.\frac{\delta\Gamma_i}{\delta A_\mu}\right|_{\chi_o} + \left.\frac{\delta\Gamma_i}{\delta\chi_{jo}}\right|_A \left.\frac{\delta\chi_{io}}{\delta A_\mu}\right|_\lambda = \Gamma_{i;\mu} + \Gamma_{ij}\mathcal{G}_{j;\mu}. \tag{9.127}$$

So we obtain

$$\Gamma_{i;\mu} = -\Gamma_{ii'}\mathcal{G}_{i';\mu} \qquad \mathcal{G}_{i;\mu} = -\mathcal{G}_{ii'}\Gamma_{i';\mu}. \tag{9.128}$$

Eq. (9.128) states that the composite operator $\Gamma_{i;\mu}$ is minus the 1-particle irreducible part of the corresponding Green's function.

We consider now the 3-legs composite Green's functions by differentiating Eq. (9.117) with respect to A_μ at constant λ:

$$\begin{aligned}
0 &= \left.\frac{\delta\Gamma_{ij}}{\delta A_\mu}\right|_\lambda \mathcal{G}_{jm} + \Gamma_{ij}\mathcal{G}_{jm;\mu} \\
&= \left[\Gamma_{ij;\mu} + \Gamma_{ijn}\left.\frac{\delta\chi_{no}}{\delta A_\mu}\right|_\lambda\right]\mathcal{G}_{jm} + \Gamma_{ij}\mathcal{G}_{jm;\mu} \\
&= \left[\Gamma_{ij;\mu} + \Gamma_{ijn}\mathcal{G}_{n;\mu}\right]\mathcal{G}_{jm} + \Gamma_{ij}\mathcal{G}_{jm;\mu},
\end{aligned} \tag{9.129}$$

so that:

$$\mathcal{G}_{ij;\mu} = -\mathcal{G}_{ii'}\mathcal{G}_{jj'}\left[\Gamma_{i'j';\mu} + \Gamma_{i'j'n}\mathcal{G}_{n;\mu}\right] \tag{9.130}$$

and by means of Eq. (9.128) we obtain

$$\mathcal{G}_{ij;\mu} = -\mathcal{G}_{ii'}\mathcal{G}_{jj'}\left[\Gamma_{i'j';\mu} - \Gamma_{i'j'n}\mathcal{G}_{nn'}\Gamma_{n';\mu}\right]. \tag{9.131}$$

The meaning of (9.131) is clear. In fact, it guaranties that the Vertex part is 1-particle irreducible also along the direction of the composite operator. Note that when the symmetry is broken we can have propagation of single particle lines that becomes composite line, but (9.131) tells us that we must discard these contributions to the vertex function.

We consider now the 4-legs Green's function by differentiating Eq. (9.122) with respect to A_μ:

$$\begin{aligned}
\mathcal{G}_{ijm;\mu} &= -\mathcal{G}_{ii';\mu}\mathcal{G}_{jj'}\mathcal{G}_{mm'}\Gamma_{i'j'm'} - \mathcal{G}_{ii'}\mathcal{G}_{jj';\mu}\mathcal{G}_{mm'}\Gamma_{i'j'm'} - \mathcal{G}_{ii'}\mathcal{G}_{jj'}\mathcal{G}_{mm';\mu}\Gamma_{i'j'm'} \\
&\quad -\mathcal{G}_{ii'}\mathcal{G}_{jj'}\mathcal{G}_{mm'}\left[\Gamma_{i'j'm';\mu} + \Gamma_{i'j'm'n}\mathcal{G}_{n;\mu}\right]
\end{aligned} \tag{9.132}$$

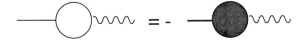

Figure 9.3: Some examples of the equations for the composite operators. Full bubbles are vertex parts while void ones are Green's functions.

So also $\Gamma_{ijk;\mu}$ is 1-particle irreducible.

In Fig. 9.3 we show the diagrammatic representation of Eq. (9.127) and Eq. (9.130).

924   Perturbation Theory

At this point we can build up the perturbation theory. We regard the quadratic part as the free part and consider the reminders as perturbation. The results obtained in 9.2.3 for the vertex functions guaranties that they are the 1-particle irreducible part of the Green's function. In this way we have an operative prescription to compute them. We begin by introducing the following four propagators:

$$\begin{pmatrix} \mathcal{G}_{11}^o(k) & \mathcal{G}_{12}^o(k) \\ \mathcal{G}_{21}^o(k) & \mathcal{G}_{22}^o(k) \end{pmatrix} = \begin{pmatrix} v_{22} + u_{22}\omega^2 + z_{22}\mathbf{k}^2 & -v_{12} - \omega w_{12} \\ -v_{12} + \omega w_{12} & v_{11} + z_{11}\mathbf{k}^2 \end{pmatrix} \frac{1}{D(k)} \quad (9.133)$$

where:

$$D(k) = v_{11}v_{22} - v_{12}^2 + (u_{22}v_{11} + w_{12}^2)\omega^2 + (v_{11}z_{22} + v_{22}z_{11} + z_{11}z_{22}\mathbf{k}^2)\mathbf{k}^2 + z_{11}u_{22}\omega^2\mathbf{k}^2 \quad (9.134)$$

$$\begin{array}{rcll}
(k,1)\text{------------}(-k,1) & = & \mathcal{G}_{11}^o(k) & (9.135) \\
(k,2)\text{_____}(-k,2) & = & \mathcal{G}_{22}^o(k) & (9.136) \\
(k,1)\text{-----_____}(-k,2) & = & \mathcal{G}_{12}^o(k) & (9.137) \\
(k,2)\text{_____-----}(-k,1) & = & \mathcal{G}_{21}^o(k) & (9.138)
\end{array}$$

the non irrelevant interaction vertices treated as perturbations are the following ones:

$$\{v_1, v_2, v_{222}, v_{122}, v_{2222}\} \; . \quad (9.139)$$

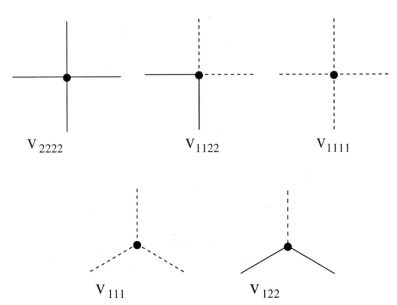

Figure 9.4: Interaction vertices corresponding to non-vanishing bare coupling.

The perturbation theory is the usual one. We have to draw all possible diagrams, associate to each line the corresponding Green's function, and to each vertex its corresponding running coupling. Then we have to compute the symmetry factor according to the standard rules (see for example Ref. [98]), and perform the integral over the internal momenta. In Fig. (9.4) we show some of the interaction vertices of this theory.

The Ward identities that we have found hold exactly, so if we perform a loop expansion of the vertex functions, Ward identities will hold order by order (see also for a detailed discussion 10.1). So we will use this property in 10.2 to impose some restrictions on the bare and renormalized running couplings.

925   Perturbation Theory for Composite Greens Functions

We have seen that to compute the vertex parts we must compute the Green's function, so we write their explicit expressions. For example, we consider $\mathcal{G}_{i;\mu}$:

$$
\begin{aligned}
\mathcal{G}_{i;\mu}(x,y) &\equiv \frac{\delta F}{\delta \lambda_i(x) \delta A_\mu(y)} \\
&= \frac{\delta}{\lambda_i(x)} \langle S_{;\mu}(y) \rangle \\
&= \langle S_{i;\mu}(x,y) \rangle - \langle S_i(x) \rangle \langle S_{;\mu}(y) \rangle + \langle S_{;\mu}(x) \rangle \langle S_i(y) \rangle \\
&= \langle S_{i;\mu}(x,y) \rangle_{Connected}, \tag{9.140}
\end{aligned}
$$

where we defined as usual $S_i(x) = \delta S / \delta \lambda_i(x)$. In a similar way we can obtain the definition of the other Green's functions. So we only need the various derivatives of S [cf. (8.9)]:

$$
\begin{align}
S_1(x) &= \chi_{1o} + \tilde{\chi}_1(x) \tag{9.141} \\
S_2(x) &= \tilde{\chi}_2(x) \tag{9.142} \\
S_{;0}(x) &= \chi_{1o}^2 + 2\chi_{1o}\tilde{\chi}_1(x) + \tilde{\chi}_1^2(x) + \tilde{\chi}_2^2(x) \tag{9.143} \\
S_{;j}(x) &= 2\chi_{1o}(\nabla)_j\tilde{\chi}_2(x) + 2\tilde{\chi}_1(x)(\nabla)_j\tilde{\chi}_2(x) - 2\tilde{\chi}_2(x)(\nabla)_j\tilde{\chi}_1(x) \\
&\quad -2A_j(x)\left[\chi_{1o}^2 + 2\chi_{1o}\tilde{\chi}_1(x) + \tilde{\chi}_1^2(x) + \tilde{\chi}_2^2(x)\right] \tag{9.144} \\
S_{;jl}(x,y) &= -2\delta_{jl}\delta(x-y)\left[\chi_{1o}^2 + 2\chi_{1o}\tilde{\chi}_1(x) + \tilde{\chi}_1^2(x) + \tilde{\chi}_2^2(x)\right]. \tag{9.145}
\end{align}
$$

Some comments are in order. Note that these equations are correct only if the true value χ_{1o} does not depend on λ and A_μ (otherwise we should consider also their dependence in the derivatives). We can simply define χ_{1o} as the value of $\langle\chi_1\rangle$ for $\lambda_i = 0$ and $A_\mu = 0$; in this sense the notation may be misleading because so far we have indicated with χ_{1o} the average with finite external sources. Note also that the constant part of this function will give contribution to the *connected* Green's function only if we have a single external leg, for example the total density [99]:

$$
\mathcal{G}_{;0}(x) = \langle S_{;0}(x)\rangle = \chi_{1o}^2 + 2\chi_{1o}\langle\tilde{\chi}_1(x)\rangle + \langle\chi_1(x)^2 + \chi_2(x)^2\rangle = n. \tag{9.146}
$$

Insertions

From the above equations we can find the composite insertions. We define the insertions as $\gamma_{i;\mu}$ and $\gamma_{ij;\mu}$ and we can obtain their expressions from (9.143) and (9.144) by Fourier transform ($k \neq 0$):

$$
S_{;0}(k) = 2\frac{\chi_{1o}}{\sqrt{\beta\Omega}}\tilde{\chi}_1(k) + \frac{1}{\beta\Omega}\sum_{k_1,k_2}\delta_{k_1+k_2-k}\left[\tilde{\chi}_1(k_1)\tilde{\chi}_1(k_2) + \tilde{\chi}_2(k_1)\tilde{\chi}_2(k_2)\right],
$$
$$\tag{9.147}$$

and for the spatial part ($\mathbf{A} = 0$ in the following term):

$$
S_{;j}(k) = 2ik_j\chi_{1o}\tilde{\chi}_2(k) + 2i\sum_{k_1,k_2}\delta_{k_1+k_2-k}(\mathbf{k}_2 - \mathbf{k}_1)_j\tilde{\chi}_1(k_1)\tilde{\chi}_2(k_2). \tag{9.148}
$$

The second derivative with respect to A_l gives:

$$
S_{;jl}(k) = -2\delta_{jl}S_{;0}(k). \tag{9.149}
$$

From (9.147) we can read $\gamma_{i;0}$ and $\gamma_{ij;0}$:

$$
\begin{align}
\gamma_{i;0}(k) &= 2\chi_{1o}\delta_{i,1} \tag{9.150} \\
\gamma_{ij;0}(k_1,k_2) &= 2\delta_{i,j}. \tag{9.151}
\end{align}
$$

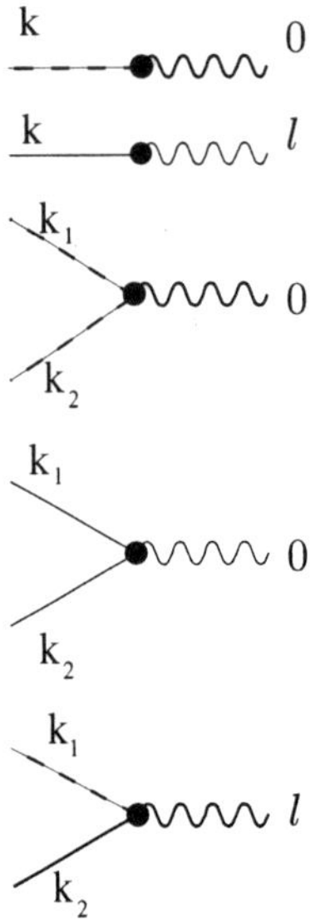

$$
\begin{aligned}
&\overset{k}{\text{------}\bullet\hspace{-0.3em}\text{WW}}\ 0 \ = \ \gamma_{1;0}(k) && = \ 2\chi_{1o} \\[1em]
&\overset{k}{\text{------}\bullet\hspace{-0.3em}\text{WW}}\ l \ = \ \gamma_{2;l}(k) && = \ -2i\mathbf{k}_l\chi_{1o}
\end{aligned}
$$

$$
\gamma_{11;0}(k_1,k_2) \ = \ 2
$$

$$
\gamma_{22;0}(k_1,k_2) \ = \ 2 \tag{9.156}
$$

$$
\gamma_{12;l}(k_1,k_2) \ \cdot= \ -2i(\mathbf{k}_2-\mathbf{k}_1)_l
$$

Figure 9.5: Summary of the composite insertions

We define the γ functions with the symmetry factors given by the interaction vertex so that we can use the same perturbation theory. In this way we see that the 2 in (9.151) comes from the symmetry of the term χ_i^2 (2!). From (9.148) we read $\gamma_{i;l}$ and $\gamma_{ij;l}$:

$$
\gamma_{i;l}(k) \ = \ -2i\mathbf{k}_l\chi_{1o}\delta_{i,2}, \tag{9.152}
$$

$$
\gamma_{ij;l}(k_1,k_2) \ = \ \begin{pmatrix} 0 & -2i(\mathbf{k}_2-\mathbf{k}_1)_l \\ 2i(\mathbf{k}_2-\mathbf{k}_1)_l & 0 \end{pmatrix}, \tag{9.153}
$$

and finally the second derivatives:

$$
\gamma_{i;lm}(k_1;k_2) \ = \ -2\chi_{1o}\delta_{lm} \tag{9.154}
$$

$$
\gamma_{ij;lm}(k_1,k_2;k_3,-k_1-k_2-k_3) \ = \ -2\delta_{ij}\delta_{lm}. \tag{9.155}
$$

Recall that when we calculate the correlation functions we must pair the external $\tilde{\chi}(k)$ with the fields coming from the action or the fields coming from the composite parts written above; in that case the external momentum will be minus the momentum of the composite part. This is the origin of the minus sign in (9.152) and (9.153).

In the following we will consider the lowest order calculation for the principal Green's function to check the correctness of our results for γ.

1-Particle

We consider now the real composite correlation function:

$$\mathcal{G}_{i;0}(x,y) = \left\langle \tilde{\chi}_i(x)[2\chi_{1o}\tilde{\chi}_1(y) + \tilde{\chi}_1(y)^2 + \tilde{\chi}_2(y)^2] \right\rangle_C$$
$$= 2\chi_{1o}\mathcal{G}_{i1}(x,y) + \left\langle \tilde{\chi}_i(x)[\tilde{\chi}_1(y)^2 + \tilde{\chi}_2(y)^2] \right\rangle_C . \qquad (9.157)$$

We can write the lowest order of this Green's function:

$$\mathcal{G}_{i;0}^{(0)}(x,y) = 2\chi_{1o}\mathcal{G}_{i1}^{(0)}(x,y) \qquad \Rightarrow \qquad \mathcal{G}_{i;0}^{(0)}(k) = 2\chi_{1o}\mathcal{G}_{i1}^{(0)}(k). \qquad (9.158)$$

Once we have eliminated the external Green's functions (see (9.128)) we obtain (*i.e.*, the rest is at least of the order of one loop):

$$\begin{cases} \Gamma_{1;0}^{(0)} = -\gamma_{1;0}(k,-k) = -2\chi_{1o} \\ \Gamma_{2;0}^{(0)} = -\gamma_{2;0}(k,-k) = 0. \end{cases} \qquad (9.159)$$

In a similar way we write ($\mathbf{A} = 0$):

$$\mathcal{G}_{i;l}(x,y) = 2\left\langle \tilde{\chi}_i(x)\left[\chi_{1o}(\nabla_y)_l\tilde{\chi}_2(y) + \tilde{\chi}_1(y)(\nabla_y)_l\tilde{\chi}_2(y) - \tilde{\chi}_2(y)(\nabla_y)_l\tilde{\chi}_1(y)\right]\right\rangle_C$$
$$= 2\chi_{1o}(\nabla_y)_l\mathcal{G}_{i2}(x,y) + 2\left\langle \tilde{\chi}_i(x)\left[\tilde{\chi}_1(y)(\nabla_y)_l\tilde{\chi}_2(y) - \tilde{\chi}_2(y)(\nabla_y)_l\tilde{\chi}_1(y)\right]\right\rangle_C , \qquad (9.160)$$

so we obtain the lowest-order expression:

$$\mathcal{G}_{i;l}^{(0)}(k) = 2\chi_{1o}(-ik_l)\mathcal{G}_{i2}^{(0)}(k) \qquad (9.161)$$

$$\mathcal{G}_{i;l}^{(0)}(k) = -\mathcal{G}_{ii'}^{(0)}\Gamma_{i';l}^{(0)}(k), \qquad (9.162)$$

and the lowest-order vertex parts are:

$$\begin{cases} \Gamma_{1;l}^{(0)}(k) = -\gamma_{1;l}(k,-k) = 0 \\ \Gamma_{2;l}^{(0)}(k) = -\gamma_{2;l}(k,-k) = 2ik_l\chi_{1o}. \end{cases} \qquad (9.163)$$

2-Particle

In a similar way we consider now the composite Green's functions with two external particle lines:

$$\mathcal{G}_{ij;0}(x,y;z) = 2\chi_{1o}\left\langle \tilde{\chi}_i(x)\tilde{\chi}_j(y)\tilde{\chi}_1(z)\right\rangle_C$$
$$+ \left\langle \tilde{\chi}_i(x)\tilde{\chi}_j(y)\tilde{\chi}_1(z)^2 \right\rangle_C + \left\langle \tilde{\chi}_i(x)\tilde{\chi}_j(y)\tilde{\chi}_2(z)^2 \right\rangle_C , \qquad (9.164)$$

this gives for the lowest order vertex part the following expression:

$$\Gamma_{ij;0}^{(0)}(k_1,k_2;-k_1-k_2) = -\gamma_{ij;0}(k_1,k_2;-k_1-k_2) . \qquad (9.165)$$

The current part gives instead:

$$\mathcal{G}_{ij;l}(x,y,z) = 2\chi_{1o}(\nabla_z)_l \mathcal{G}_{ij2}(x,y,z)$$
$$+2\left\langle \tilde{\chi}_i(x)\tilde{\chi}_j(y)\left[\tilde{\chi}_1(z)(\nabla_z)_l\tilde{\chi}_2(z) - \tilde{\chi}_2(z)(\nabla_z)_l\tilde{\chi}_1(z)\right]\right\rangle_C, \quad (9.166)$$

and so for the vertex part we get:

$$\Gamma_{ij;l}(k_1,k_2;-k_1-k_2) = -\gamma_{ij;l}(k_1,k_2;-k_1-k_2). \quad (9.167)$$

926   Symmetry Limitations on γ

We take now into account the fact that the system is invariant under: rotation, translation, Parity, and Time-reversal. In this way we can pose conditions on the k-dependence of the insertions:

$$\begin{cases} \gamma_{1;0}(k) &= \bar{\gamma}_{2;0}(\omega^2,\mathbf{k}^2) \\ \gamma_{2;0}(k) &= \omega\bar{\gamma}_{2;0}(\omega^2,\mathbf{k}^2) \\ \gamma_{1;l}(k) &= i\mathbf{k}_l\omega\bar{\gamma}_{1;v}(\omega^2,\mathbf{k}^2) \\ \gamma_{2;l}(k) &= i\mathbf{k}_l\bar{\gamma}_{2;v}(\omega^2,\mathbf{k}^2) \end{cases} \quad (9.168)$$

see (9.107). And also (see (9.108))

$$\begin{cases} \gamma_{ii;0}(k_1,k_2) &= \bar{\gamma}_{ii;0}(k_1,k_2) \\ \gamma_{12;0}(k_1,k_2) &= (\omega_1-\omega_2)\bar{\gamma}_{12;0}(k_1,k_2) \\ \gamma_{ii;l}(k_1,k_2) &= i(\mathbf{k}_1+\mathbf{k}_2)_l(\omega_1+\omega_2)\bar{\gamma}_{ii;v}(k_1,k_2) \\ \gamma_{12;l}(k_1,k_2) &= i(\mathbf{k}_1-\mathbf{k}_2)_l\bar{\gamma}_{12;v}(k_1,k_2), \end{cases} \quad (9.169)$$

The bare values for these running couplings are the following:

$$\begin{cases} \bar{\gamma}_{1;0} &= 2\chi_{1o} \\ \bar{\gamma}_{2;0} &= 0 \\ \bar{\gamma}_{1;v} &= 0 \\ \bar{\gamma}_{2;v} &= -2\chi_{1o} \end{cases} \quad \begin{cases} \bar{\gamma}_{ii;0} &= 2 \\ \bar{\gamma}_{12;0} &= 0 \\ \bar{\gamma}_{ii;v} &= 0 \\ \bar{\gamma}_{12;v} &= 2 \\ \bar{\gamma}_{21;v} &= -2 \end{cases} \quad (9.170)$$

Chapter 10

Use of Ward Identities

10.1 Comment on Ward Identities and the Loop Expansion

In this Section we ask what happens to the Ward identities when we introduce a parameter λ (not to be confused with the source $\lambda_i(x)$) in the definition of F. The purpose is to set up a loop expansion similar to the standard one [98]. The new Free energy is defined as follows:

$$F_\lambda \equiv \ln \int \mathcal{D}\chi\, e^{\frac{S}{\lambda}}, \qquad (10.1)$$

from this definition we define all connected Green's functions of the theory, $\mathcal{G}^\lambda$, by performing the functional derivative with respect to the sources $\lambda_i(x)$ or $A_\mu(x)$:

$$\mathcal{G}_1^\lambda(x) \equiv \frac{\delta F_\lambda}{\delta\lambda(x)} \qquad (10.2)$$

$$\mathcal{G}_2^\lambda(x,y) \equiv \frac{\delta^{(2)} F_\lambda}{\delta\lambda(x)\delta\lambda(y)} \qquad (10.3)$$

$$\mathcal{G}_3^\lambda(x,y,z) \equiv \frac{\delta^{(3)} F_\lambda}{\delta\lambda(x)\delta\lambda(y)\delta\lambda(y)} \qquad (10.4)$$

$$\cdots$$

$$\mathcal{G}_{1;1}^\lambda(x;y) \equiv \frac{\delta^{(2)} F_\lambda}{\delta\lambda(x) A_\mu(y)} \qquad (10.5)$$

$$\cdots$$

So in the same way we can define the Legendre transform Γ^λ and all the vertex parts as functional derivatives of Γ^λ. Now we ask which is the connection of our new correlation functions with the correlation functions that we can calculate with perturbative methods. We define, in particular, the following functions:

$$\begin{cases} \mathcal{G}_2^P &= \langle \tilde{\chi}_i \tilde{\chi}_j \rangle_\lambda \\ \mathcal{G}_{1;1}^P &= \langle \tilde{\chi}_i J_\mu \rangle_\lambda \\ \dots \end{cases} \qquad (10.6)$$

where

$$\langle A \rangle_\lambda \equiv \frac{\int \mathcal{D}\chi\, e^{S/\lambda} A}{\int \mathcal{D}\chi\, e^{S/\lambda}}. \qquad (10.7)$$

For the "perturbative" correlation functions we can easily develop a power series expansion in λ. The connection of the perturbative and the λ correlation functions is straightforward:

$$\mathcal{G}_1^\lambda \equiv \frac{\delta F}{\delta \lambda_i} = \frac{\langle \chi \rangle_\lambda}{\lambda} = \frac{\mathcal{G}_1^P}{\lambda} \qquad \Rightarrow \qquad \chi_1^\lambda = \frac{\chi_1^P}{\lambda} \qquad (10.8)$$

$$\mathcal{G}_2^\lambda = \frac{\langle \chi\chi \rangle_\lambda}{\lambda^2} = \frac{\mathcal{G}_2^P}{\lambda^2} \qquad (10.9)$$

$$\mathcal{G}_N^\lambda = = \frac{\mathcal{G}_N^P}{\lambda^N} \qquad (10.10)$$

$$\mathcal{G}_{1;1}^\lambda \equiv \frac{\delta^{(2)} F_\lambda}{\delta \lambda_i A_\mu} = \frac{\langle \chi \rangle_\lambda}{\lambda^2} = \frac{\mathcal{G}_{1;1}^P}{\lambda^2} \qquad (10.11)$$

$$\mathcal{G}_{N;M}^\lambda = = \frac{\mathcal{G}_{N;M}^P}{\lambda^{N+M}} \qquad (10.12)$$

We define the perturbative vertex parts (symbolically):

$$\mathcal{G}_N^P = -(\mathcal{G}_2^P)^N \Gamma_N^P \qquad (10.13)$$

$$\mathcal{G}_{N;M}^P = -(\mathcal{G}_2^P)^N \Gamma_{N;M}^P, \qquad (10.14)$$

so that we have (keeping the same definitions of the λ functions):

$$\Gamma_N^\lambda = \lambda^N \Gamma_N^P \qquad (10.15)$$

$$\Gamma_{N;M}^\lambda = \lambda^{N-M} \Gamma_{N;M}^P. \qquad (10.16)$$

The Ward identities are correct for the λ Green's functions, but *a priori* not for the perturbative ones. So we write how they change when written in terms of the perturbative vertex parts [cf. (8.61):

$$\Gamma_2^\lambda \chi_{1o}^\lambda + \Gamma_1^\lambda - k\Gamma_{1;1}^\lambda = 0, \qquad (10.17)$$

this expression becomes:

$$\lambda[\Gamma_2^P \chi_{1o}^P + \Gamma_1^P] - k\Gamma_{1;1}^P = 0., \qquad (10.18)$$

and similarly for the other identities. We can thus conclude that the identities are valid order by order in λ only if we multiply by λ the non composite part of the identity.

This result guarantees that the lowest order vertex part of the composite operators is of the same order (in the loop expansion) of the lowest order vertex part of the non composite operators.

10.2 Ward Identities in the Loop Expansion

In this Section we implement the gauge symmetry in the theory by means of Ward identities. This is done by considering the constraints that they give on the running couplings. Note that the following equations have been derived for the bare coupling, but by expanding them in power series of the renormalized (physical) couplings it is easy to show that the same relations hold for the renormalized couplings too. The origin of this fact stems on the gauge invariance of the renormalized theory.

1021 Non Singular Ward Identities

We begin by writing the lowest-order expressions for the vertex functions:

$$
\begin{aligned}
\bar{\Gamma}_1(0) &= -v_1 \\
\bar{\Gamma}_2(0) &= -v_2 \\
\bar{\Gamma}_{11}(k) &= v_{11} + z_{11}\mathbf{k}^2 \\
\bar{\Gamma}_{22}(k) &= v_{22} + u_{22}\omega^2 + z_{22}k^2 \\
\bar{\Gamma}_{12}(k) &= v_{12} + \omega w_{12} \\
\bar{\Gamma}_{21}(k) &= v_{12} - \omega w_{12} \\
\bar{\Gamma}_{222}(k_1, k_2) &= v_{222} \\
\bar{\Gamma}_{122}(k_1, k_2) &= v_{122} \\
\bar{\Gamma}_{2222}(k_1, k_2) &= v_{2222}
\end{aligned}
\tag{10.19}
$$

and for the composite parts:

$$
\begin{aligned}
\bar{\Gamma}_{1;0}(k) &= -\gamma_{1;0}(k) & &= -2\chi_{1o} \\
\bar{\Gamma}_{2;0}(k) &= -\omega\bar{\gamma}_{2;0}(k) & &= 0 \\
\bar{\Gamma}_{1;l}(k) &= -i\mathbf{k}_l\omega\bar{\gamma}_{1;v}(k) & &= 0 \\
\bar{\Gamma}_{2;l}(k) &= -i\mathbf{k}_l\bar{\gamma}_{2;v}(k) & &= 2i\mathbf{k}_l\chi_{1o} \\
\bar{\Gamma}_{ii;0}(k_1, k_2;) &= -\gamma_{ii;0}(k_1, k_2) & &= -2 \\
\bar{\Gamma}_{12;0}(k_1, k_2) &= (\omega_1 - \omega_2)\bar{\gamma}_{12;0}(k_1, k_2) & &= 0 \\
\bar{\Gamma}_{ii;l}(k_1, k_2) &= -i(\mathbf{k}_1 + \mathbf{k}_2)_l(\omega_1 + \omega_2)\bar{\gamma}_{ii;v}(k_1, k_2) & &= 0 \\
\bar{\Gamma}_{12;l}(k_1, k_2) &= -i(\mathbf{k}_1 - \mathbf{k}_2)_l\bar{\gamma}_{12;v}(k_1, k_2) & &= -2i(\mathbf{k}_1 - \mathbf{k}_2)_l
\end{aligned}
\tag{10.20}
$$

The third column indicates the bare values of the insertions.

We can soon set v_1 and v_2 to zero from the WI (8.72) and the condition that the external source are let to vanish ($v_1 = \lambda_1$ and $v_2 = \lambda_2$).

One other simple case is the v_{12} coupling, in fact (8.73) can be used at $k_1 = 0$. We can do this because $\bar{\Gamma}_{21}$ is not diverging for $k \to 0$ due to its positive IR dimension, moreover the composite operator $\bar{\Gamma}_{1;v}$ has at least dimension 0, and with the factor k in front becomes regular in the IR (it vanishes). In this way we can write:

$$
\bar{\Gamma}_{21}(0)\chi_{1o} + \bar{\Gamma}_2(0) = 0 \qquad \Rightarrow \qquad \Gamma_{21}(0) = \Gamma_{12}(0) = v_{12} = 0 .
\tag{10.21}
$$

In a similar way we can eliminate the coupling v_{22} (see (8.73)):

$$\Gamma_{22}(0) = v_{22} = 0. \qquad (10.22)$$

The last simple case is represented by v_{222}, in fact (8.77) can be evaluated at $k = 0$ because it is free from IR divergences by power counting:

$$\bar{\Gamma}_{222}(0,0) = v_{222} = 0. \qquad (10.23)$$

With the help of the above equations and the relevance of the running couplings we can simplify greatly the list of retained running couplings given in (9.56). We begin by eliminating the irrelevant ones. In this way we are left with the following list (v_0 is the Free energy and does not appear in the perturbation theory):

$$\{v_1, v_2, v_{11}, v_{12}, v_{22}, w_{12}, u_{22}, z_{22}, v_{122}, v_{222}, v_{2222}\}. \qquad (10.24)$$

now we use the fact that v_1, v_2, v_{12}, v_{22}, and v_{222} vanish. In this way we are left with:

$$\{v_{11}, w_{12}, u_{22}, z_{22}, v_{122}, v_{2222}\}. \qquad (10.25)$$

These running coupling are all dimensionless for $d = 3$ and for $d < 3$ the last two acquire a positive dimension: $\epsilon/2$ and ϵ respectively. It is possible at this point to treat the problem within an epsilon expansion around the critical dimension $d = 3$ where the most relevant running couplings are dimensionless. In the following we will see how the problem can be simplified more by using other Ward identities, but a comment on the scaling of the condensate density is necessary at this point. As a matter of fact the Ward identities depend on the *exact* condensate density, because they are obtained from the generating functional $\Gamma[\chi_o, A_\nu]$ which is a function of the exact condensate density. A subtle point is that the value of this condensate density is not necessarily at zero external fields λ_i, but in general it corresponds to an unknown value of the external sources when the other parameters are fixed. So the equality:

$$\Gamma_1(k = 0, v_1) = \lambda_1 \qquad (10.26)$$

must be regarded as an equation that defines the physical condensate density at fixed λ_1 and for $\lambda_1 = 0$ it constitutes an operative definition for the condensate density (also known as the tadpole equation). So v_1 constitutes the bare source that should be found as a power series of the bare coupling and used to fix the physical source $\lambda = 0$. Note that by power counting this equation is regular in the IR, so that we can invert it in terms of the bare running couplings. The lowest order of v_1 is clearly zero, giving a gapless approximation for the system and no contribution at one-loop. Higher order terms comes to eliminate the "mass" renormalization diagrams appearing in the $\mathcal{G}_{22}$ propagators and can be neglected consistently in the calculation by neglecting the "mass" renormalization diagrams. The exact constraints found in the following on the IR behavior of the $\mathcal{G}_{22}$ guarantees that no extra singularities can be introduced by this equation at two-loops or beyond.

1022   Singular LowestOrder Ward Identities

In the following we consider the singular (in the IR) Ward identities at zero loop.

Ward Identities of 2-Point with 1-Point Vertex Functions

From (8.73)

$$\begin{cases} \bar{\Gamma}_{21}(k)\chi_{1o} - i(i\omega)\bar{\Gamma}_{1;0}(-k) - ik_i\bar{\Gamma}_{1,i}(-k) &= 0 \\ \bar{\Gamma}_{22}(k)\chi_{1o} - i(i\omega)\bar{\Gamma}_{2;0}(-k) - ik_i\bar{\Gamma}_{2,i}(-k) &= 0 \end{cases} \tag{10.27}$$

Substituting (10.19) and (10.20) we have:

$$\begin{cases} -\omega[w_{12}\chi_{1o} + \gamma_{1;0}(-k)] - \omega\mathbf{k}^2\bar{\gamma}_{1;v}(-k) &= 0 \\ \omega^2[u_{22}\chi_{1o} + \bar{\gamma}_{2;0}(-k)] + [z_{22}\chi_{1o} + \bar{\gamma}_{2;v}(-k)]\mathbf{k}^2 &= 0, \end{cases} \tag{10.28}$$

so the final result for the running couplings at the lowest order is:

$$\begin{cases} w_{12}\chi_{1o} &= -\gamma_{1;0} \\ \bar{\gamma}_{1;v} &= 0 \\ \bar{\gamma}_{2;0} &= \chi_{1o}u_{22} \\ z_{22}\chi_{1o} + \bar{\gamma}_{2;v} &= 0 \end{cases} \tag{10.29}$$

These identities are consistent with the starting values of the running couplings, as can be seen by comparison with (10.20).

Ward Identities of 3-Point with 2-Point Vertex Functions

From (8.75)

$$\bar{\Gamma}_{221}(k_1, k_2)\chi_{1o} + \bar{\Gamma}_{22}(-k_2) - \bar{\Gamma}_{11}(k_1 + k_2)$$
$$-i(i\omega_1)\bar{\Gamma}_{21;0}(k_2, -k_1 - k_2) - ik_l^1\bar{\Gamma}_{21;l}(k_2, -k_1 - k_2) = 0 \tag{10.30}$$

$$\begin{aligned} 0 &= v_{122}\chi_{1o} - v_{11} + u_{22}\omega_2^2 + \bar{\gamma}_{12;0}[\omega_1^2 + 2\omega_1\omega_2] \\ &\quad + \mathbf{k}_2^2[z_{22} - z_{11}] + [\mathbf{k}_1^2 + 2\mathbf{k}_1\mathbf{k}_2][\bar{\gamma}_{12;v} - z_{11}] \end{aligned} \tag{10.31}$$

This gives for the lowest-order couplings:

$$\begin{cases} v_{11} &= \chi_{1o}v_{122} \\ u_{22} &= 0 \\ z_{22} &= z_{11} \\ z_{11} &= -\bar{\gamma}_{21;v} \\ 0 &= \bar{\gamma}_{12;0} \end{cases} \tag{10.32}$$

Again, as a simple check, we see that at the lowest order the Ward identities are valid for the bare couplings. Note, however, that the above form of the WI is not so useful to establish constraints on the derivatives of the Γ functions, this is due to the fact that the running coupling for some derivatives does not appear at the lowest order (*i.e.* has no insertion in the bare action). This is the case of u_{22} that from this WI seems vanish identically, while from the previous one is connected with $\bar{\gamma}_{2;0}$. To obtain the correct constraints on the derivatives we will consider the expansion of the complete vertex functions in Section 10.3.

Ward Identities of 4-Point with 3-Point Vertex Functions

From (8.78)

$$\bar{\Gamma}_{2222}(k_1, k_2, k_3)\chi_{1o} - \bar{\Gamma}_{122}(-k_2 - k_3, k_2) - \bar{\Gamma}_{122}(k_1 + k_3, k_2)$$
$$-\bar{\Gamma}_{122}(k_1 + k_2, k_3) - i(i\omega_1)\bar{\Gamma}_{222;0}(k_2, k_3, -k_1 - k_2 - k_3)$$
$$-ik_l^1\bar{\Gamma}_{222;l}(k_2, k_3, -k_1 - k_2 - k_3) = 0 \qquad (10.33)$$

At the lowest order we have:

$$\chi_{1o}v_{2222} - 3v_{122} - \omega\gamma_{222;0} - i\mathbf{k}_l^1\gamma_{222;l}(k_2, k_3, -k_1 - k_2 - k_3) = 0 \qquad (10.34)$$

So that we get the following conditions for the running couplings:

$$\begin{cases} v_{2222}\chi_{1o} &=& 3v_{122} \\ \gamma_{222;0} &=& 0 \\ \gamma_{222;l} &=& 0 \end{cases} \qquad (10.35)$$

Now we can use the relations found for $v_{122} = v_{11}/\chi_{1o}$ and

$$v_{2222} = 3v_{122}/\chi_{1o} = 3v_{11}/\chi_{1o}^2$$

to reduce the number of running coupling appearing in (10.25) to only four:

$$\{v_{11}, w_{12}, u_{22}, z_{22}\}. \qquad (10.36)$$

In the following we need to write Renormalization Group equations for only these four running coupling.

10.3 Exact Constraints from Local Part of WI

In this Section we investigate the constraints given by the local part of WI on the momentum derivatives of vertex functions. The first study of this kind was performed by Gavoret and Nozières [82]. Similar results with different approaches were then obtained in Ref. [31] and Ref. [83]. As we show in the following Sections, in the χ-representation the same results are much simpler to obtain, and we will see that they lead to important constraints for the RG equations.

1031   Loop Expansion and Divergences in the Pertur bation Theory

Before studying the local part of the WI we need to analyze in some details the appearance of singularities in the perturbation theory. We have reduced the interaction vertices to only two: v_{122} and v_{2222}, it is possible to perform a simple calculation on the degree of superficial divergence $[D(E_1, E_2, n_{122}, n_{2222}, L)]$ of a diagram contributing to a Γ function with E_1 external legs of type 1, E_2 external legs of type 2, the number of interaction coupling n_{122}, n_{2222} and the order in loops L:

$$D(E_1, E_2, n_{122}, n_{2222}, L) = (d+1)L - 2n_{122} - 4n_{2222} + E_2. \qquad (10.37)$$

The simple idea is that the coupling n_{122} appears in the perturbation expansion always with "two half" of the $1/k^2$ bare propagator, so leading to a divergence of k^{-2}. In the same way the coupling v_{2222} brings "four half" of $1/k^2$ and so a divergence of k^{-4}. We must then eliminate the external divergences that have been erroneously counted by this technique by subtracting E_2.

We can now find a simple relation between the number of couplings present in the diagram and the number of internal lines I:

$$I = (3n_{122} + 4n_{2222} - E_1 - E_2)/2. \qquad (10.38)$$

The number of loops L, or of independent integrations, is given by:

$$L = I - n_{122} - n_{2222} + 1. \qquad (10.39)$$

By making use of (10.37), (10.38), and (10.39) we can write the following expression for the divergence of the diagrams:

$$D(E_1, E_2, L) = -\epsilon L - 2E_1 - E_2 + 4. \qquad (10.40)$$

We find that the degree of divergence depends only on the number of loops and not on the multiplicity of the two running couplings. This fact is very important and constitutes a necessary condition for the theory to be renormalizable. As a matter of fact the divergences can be eliminated for $\epsilon = 0$ because they depend only on the bare dimension of the vertex function, and do not change by increasing the order of the perturbation theory. This is the analogue of what happens in a ϕ^4 theory where the degree of divergence is proportional to the number of quartic interaction multiplied by ϵ, so that the theory is renormalizable for $\epsilon = 0$. In this case the role of the order in the perturbation expansion is played by the order in the expansion in loops. We can thus set up an epsilon expansion around $d = 3$.

We consider now other consequences of (10.40). From the form of the divergence the singular and regular contribution to any vertex function in

$d = 3$ can be written as follows:

$$\Gamma(k) = k^{4-2E_1-E_2} \left[\sum_{n=1}^{\infty} A_n(k) \ln^n k + B(k) \right] \qquad (10.41)$$

and for $d < 3$:

$$\Gamma(k) = k^{4-2E_1-E_2} \left[\sum_{n=1}^{\infty} A_n(k) k^{-n\epsilon} + B(k) \right] . \qquad (10.42)$$

Where $A_n(k)$ and $B(k)$ are regular functions of k^2. Note also that due to internal divergences that appear in the theory the superficial divergence appears at each loop together with the sub-leading divergences; so we can't say for instance that the 3 loops diagrams contribute only with $\ln^3 k$, but they will bring also regular, $\ln k$, and $\ln^2 k$ terms.

The definition itself of singular part is unique only for $d = 3$ because in that case no mixing between the regular functions $A_n(k)$ and the singular terms $\ln^m k$ can occur. On the contrary for $d < 3$ by expanding in power series $A_n(k)$ we obtain:

$$A_n(k) k^{-\epsilon n} = \left[A_n^{(0)} + k^2 A_n^{(1)+\cdots} \right] k^{-\epsilon n}, \qquad (10.43)$$

and if m is the order of the perturbative expansion of A_n we have problems when

$$-n\epsilon + 2m = -n'\epsilon \qquad (10.44)$$

in this case in fact we can't define in a unique way the coefficient of the singular term. This is only a formal problem, if we choose ϵ as an irrational number (10.44) can never be satisfied for any integer value of m, n, and n', so the series is well defined. In this way formally we can approach the rational values of ϵ (for example $d = 2$) as a limit over irrational values near to it.

We are now able to discuss the conditions posed by the local part of the WI. We will use the expansion (10.41) or (10.42) for the singular vertex functions: a sum of two or more vertex functions with the same bare dimensions will be always written in the form (10.41) or (10.42) (depending on the dimension). So WI become conditions on the coefficients of the singular series:

$$\sum_{n} A_n(k) k^{-n\epsilon} + B(k) = 0 \qquad \Rightarrow \qquad A_n(0) = 0, \quad B(0) = 0 \qquad (10.45)$$

we can set $k = 0$ in the expression for $A_n(k)$ and $B(k)$ because they are regular by hypothesis. This means that if one of these series is summed up to give an analytic function of $k^{-\epsilon}$, the equivalence of the coefficient guaranties that also the other linked by the WI has the same property.

One can ask what happens when we have vertex functions with different bare dimensions that appear in the same WI, for example consider the following case:

$$\sum_{n} A_n k^{-n\epsilon} + k^2 \sum_{n} A_n' k^{-n\epsilon} = 0 \qquad (10.46)$$

it is then clear that we can write:

$$A_n(k) + k^2 A'_n(k) = 0 \qquad \Rightarrow \qquad A_n(0) = 0 \tag{10.47}$$

i.e. the A' term can be neglected.

Some final comments are in order. We introduced this power series of singular terms to treat the case of finite ϵ, it is also possible to expand all the functions in ϵ

$$k^{-\epsilon} = \exp\{-\epsilon \ln k\} = \sum_n \frac{(-\epsilon \ln k)^n}{n!} \tag{10.48}$$

and obtain always power series with only logarithmic singularities. The final result must be the same, the treatment introduced clarifies a bit what happens for finite value of ϵ and will be useful to understand in a simple way how the series sums up. Another delicate point is the introduction of the irrelevant running coupling, it is in fact clear that the simple structure of the divergence is changed by the introduction of irrelevant operators. Again within the epsilon expansion these running coupling must not be taken into account for $\epsilon \to 0$. So we will discard them for the moment and will reconsider their contribution after the theory has been renormalized.

We can now consider the condition posed by the local WI part on the RG flow.

1032   Local Ward Identities

For clarity we restrict at this point the discussion to the three dimensional case. Using the concepts introduced in 10.3.1 there is no essential difference for the case $d < 3$.

We consider the following Ward identity written in terms of the symmetric functions $\mathcal{P}$ (cf. (8.73)):

$$\bar{\Gamma}_{21}(k)\chi_{1o} - i(i\omega)\bar{\Gamma}_{1;0}(-k) - ik_l\bar{\Gamma}_{1,l}(-k) = 0; \tag{10.49}$$

$$-\omega\left[\mathcal{P}_{12}(k)\chi_{1o} - \mathcal{P}_{1;0}(k) - \mathbf{k}^2\mathcal{P}_{1,v}(k)\right] = 0. \tag{10.50}$$

We know that each $\mathcal{P}$ in (10.50) has a power series expansion given by (10.41) with $4 - 2E_1 - E_2 = 0$ This means that for $k \to 0$ the leading contribution is given by the first two terms. The result must be finite so the divergences must cancel each other and by equating terms to terms in the singular series and in the regular part we get:

$$\lim_{k \to 0} \mathcal{P}_{12}(k)\chi_{1o} - \lim_{k \to 0} \mathcal{P}_{1;0}(k) = 0, \tag{10.51}$$

The last term of (10.51) can also be written in the following way:

$$\mathcal{P}_{1;0}(0) = \frac{\partial^2 \Gamma(\chi_{1o}, \mu)}{\partial \chi_{1o} \partial \mu} \frac{1}{\beta\Omega}. \tag{10.52}$$

This means that

$$\lim_{\omega \to 0} \chi_{1o} \frac{\partial \bar{\Gamma}_{12}(\omega, \mathbf{0})}{\partial \omega} = \frac{\partial^2 \Gamma(\chi_{1o}, \mu)}{\partial \chi_{1o} \partial \mu} \frac{1}{\beta \Omega}. \tag{10.53}$$

We consider now the following equation:

$$\bar{\Gamma}_{22}(k)\chi_{1o} - i(i\omega)\bar{\Gamma}_{2;0}(-k) - ik_i\bar{\Gamma}_{2,i}(-k) = 0$$
$$\mathcal{P}_{22}(k)\chi_{1o} - \omega^2 \mathcal{P}_{2;0}(k) - \mathbf{k}^2 \mathcal{P}_{2;v}(k) = 0. \tag{10.54}$$

Again in $d = 3$ we know that the dimension of $\mathcal{P}_{2;v}$ and $\mathcal{P}_{22}$ is 2 while $[\mathcal{P}_{2;0}] = 0$. This means that the last term gives no contribution in the IR. From this fact we have that $\mathcal{P}_{22}(0, \mathbf{k})/\mathbf{k}^2$ must be regular in the IR, while the divergences in the $\mathcal{P}_{22}/\omega^2$ must cancel with the divergences of $\mathcal{P}_{2;0}$. This fact is very important and will be very useful to extend the RG approach to all orders in ϵ. We can define the following function:

$$\mathcal{P}_{22}(k) = A^\omega(k)\omega^2 + A^{\mathbf{k}}(k)\mathbf{k}^2. \tag{10.55}$$

The Ward identity becomes:

$$\left[A^\omega(k)\chi_{1o} - \mathcal{P}_{2;0} \right] \omega^2 + \left[A^{\mathbf{k}}(k)\chi_{1o} - \mathcal{P}_{2;v}(k) \right] \mathbf{k}^2 = 0 \tag{10.56}$$

so again for $\mathbf{k} \to 0$ by equating the two singular series terms by terms we obtain:

$$A^{\mathbf{k}}(0) \equiv \lim_{\mathbf{k} \to 0} \frac{\partial \bar{\Gamma}_{22}(0, \mathbf{k})}{\partial \mathbf{k}^2} = \lim_{\mathbf{k} \to 0} \frac{\mathcal{P}_{2;v}(k)}{\chi_{1o}}, \tag{10.57}$$

while for $\mathbf{k} = 0$ and $\omega \to 0$:

$$A^\omega(0) \equiv \lim_{\omega \to 0} \frac{\partial \bar{\Gamma}_{22}(k)}{\partial \omega^2} = \lim_{\mathbf{k} \to 0} \frac{\mathcal{P}_{2;0}(k)}{\chi_{1o}}. \tag{10.58}$$

We must now understand what $\mathcal{P}_{2;0}$ and $\mathcal{P}_{2;v}$ really are. To this purpose we consider the basic Ward identity (8.46) and we partially differentiate with respect to A_μ (keeping χ_o constant):

$$\frac{\delta}{\delta A_\mu(y)} \left[\Gamma_i(x)\sigma_{ij}\chi_{jo}(x) - \partial_\nu \Gamma_{;\nu}(x) \right]_{\chi_o} = 0$$
$$\Gamma_{i;\mu}(x, y)\sigma_{ij}\chi_{jo}(x) - \partial_\nu^x \Gamma_{;\nu\mu}(x, y) = 0$$
$$\bar{\Gamma}_{i;\mu}(k)\sigma_{ij}\chi_{1o}\delta_{j1} - ik_\nu \bar{\Gamma}_{;\nu\mu}(k) = 0$$
$$\bar{\Gamma}_{2;\mu}(k)\chi_{1o} - ik_\nu \bar{\Gamma}_{;\nu\mu}(k) = 0. \tag{10.59}$$

From (10.59) we can write the following two equations:

$$\begin{cases} \bar{\Gamma}_{2;0}\chi_{1o} + \omega\bar{\Gamma}_{;00}(k) - ik_l\bar{\Gamma}_{;l0}(k) = 0 & (\mu = 0) \\ \bar{\Gamma}_{2;m}\chi_{1o} + \omega\bar{\Gamma}_{;0m}(k) - ik_l\bar{\Gamma}_{;lm}(k) = 0 & (\mu = m), \end{cases} \tag{10.60}$$

and in terms of the symmetric functions:

$$\left\{ \begin{array}{r} \omega \mathcal{P}_{2;0}\chi_{1o} + \omega \mathcal{P}_{;00}(k) + \mathbf{k}^2\omega \mathcal{P}_{;v0}(k) = 0 \\[2mm] i\mathbf{k}_m \mathcal{P}_{2;v}\chi_{1o} + \omega^2 i\mathbf{k}_m \mathcal{P}_{;0v}(k) - i\mathbf{k}_l \left[\dfrac{\mathbf{k}_l\mathbf{k}_m}{\mathbf{k}^2}\phi_l(k) - \left(\dfrac{\mathbf{k}_l\mathbf{k}_m}{\mathbf{k}^2} - \delta_{lm}\right)\phi_t(k) \right] = 0 \end{array} \right. \tag{10.61}$$

where we have defined:

$$\bar{\Gamma}_{;lm}(k) \equiv \frac{\mathbf{k}_l\mathbf{k}_m}{\mathbf{k}^2}\phi_l(k) + \left(\delta_{lm} - \frac{\mathbf{k}_l\mathbf{k}_m}{\mathbf{k}^2}\right)\phi_t(k)., \tag{10.62}$$

$$\left\{ \begin{array}{l} \Gamma_{;00}(k) = \mathcal{P}_{;00}(k) \\[1mm] \Gamma_{;0l}(k) = i\mathbf{k}_l\mathcal{P}_{;0v}(k) \end{array} \right. , \tag{10.63}$$

and $[\mathcal{P}_{;00}] = [\mathcal{P}_{;0v}] = 0$.

We multiply by $-i\mathbf{k}_m$ the second equation of (10.61) and sum over m:

$$\omega\left[\mathcal{P}_{2;0}\chi_{1o} + \mathcal{P}_{;00}(k) + \mathbf{k}^2\mathcal{P}_{;v0}(k)\right] = 0 \tag{10.64}$$

$$\mathbf{k}^2\left[\mathcal{P}_{2;v}\chi_{1o} + \omega^2\mathcal{P}_{;0v}(k) - \phi_l(k)\right] = 0 \tag{10.65}$$

We note that the divergences in $\mathcal{P}_{2;v}$ must cancel among themselves because $[\mathcal{P}_{;0v}] = 0$ and $[\phi_{l/t}] = 4$, while the $\ln^n(\omega^2 + \mathbf{k}^2)$ terms in $\mathcal{P}_{2;0}$ and those in $\mathcal{P}_{;00}$ cancel each other. From Eq. (10.58) we obtain:

$$\lim_{k\to 0}\mathcal{P}_{2;0}(k)\chi_{1o} = \lim_{k\to 0}-\mathcal{P}_{;00}(k) = -\left.\frac{\partial^2\Gamma(\chi_{1o},\mu)}{\partial\mu^2}\right|\frac{1}{\beta\Omega} \tag{10.66}$$

that gives for $\bar{\Gamma}_{22}$:

$$\lim_{\omega\to 0}\frac{\partial\bar{\Gamma}_{22}(\omega,\mathbf{0})}{\partial\omega^2} = -\frac{1}{\chi_{1o}^2}\left.\frac{\partial^2\Gamma(\chi_{1o},\mu)}{\partial\mu^2}\right|_{\chi_{1o}}\frac{1}{\beta\Omega} \tag{10.67}$$

Eq. (10.62) provides a definition of the superfluid density [100, 55, 101, 102] so we have (recall that the mass $m = 1/2$):

$$\phi_l(0,0) = \frac{n_s}{m} = 2n_s, \tag{10.68}$$

and:

$$\lim_{k\to 0}\frac{\partial\bar{\Gamma}_{22}(0,\mathbf{k})}{\partial\mathbf{k}^2} = \lim_{k\to 0}\frac{1}{\chi_{1o}^2}\mathcal{P}_{2;v}(k) = \frac{\phi_l(0)}{\chi_{1o}^2} = \frac{2n_s}{\chi_{1o}^2}\beta\Omega \tag{10.69}$$

In conclusion we have found a definite connection of the $k \to 0$ limit of the vertex function with thermodynamic derivatives. These connection has been done taking into account the singular nature of the vertex functions. To do this it has been very important to introduce the singular power series in the case of $\epsilon > 0$, because otherwise it is not possible to discard sub-leading terms in the WI. In fact in this case going to higher order will eliminate the overall different dimensions of the Γ functions.

It is now useful to connect these results with the Gavoret-Nozières result on the coincidence of the macroscopic with the microscopic sound velocity [82].

1033   Sound Velocity

We recall that the macroscopic sound velocity is defined as follows:

$$c^2 = \frac{n}{m} \frac{d\mu}{dn}\bigg|_{\lambda_i} = 2n / \frac{\partial n}{\partial \mu}\bigg|_{\lambda} \tag{10.70}$$

where λ are the external sources and in general are omitted in the notation because $\lambda_i = 0$ always in a physical system. In our case we perform the calculations at χ_{1o} constant so we have to invert the thermodynamic relationship to obtain c. We begin with the following equations for $\Gamma(\chi_{1o}, \mu)$:

$$\frac{\partial \Gamma(\chi_{1o}, \mu)}{\partial \chi_{1o}}\bigg|_{\mu} = \lambda_1(\chi_{1o}, \mu) \tag{10.71}$$

$$\frac{1}{\beta\Omega} \frac{\partial \Gamma(\chi_{1o}, \mu)}{\partial \mu}\bigg|_{\chi_{1o}} = -\frac{1}{\beta\Omega} \frac{\partial F(\chi_{1o}, \mu)}{\partial \mu}\bigg|_{\chi_{1o}} = -n. \tag{10.72}$$

We regard (10.71) as an implicit equation for $\chi_{1o} = \chi_{1o}(\lambda_1, \mu)$. We differentiate (10.72) with respect to μ at λ_1 constant:

$$\beta\Omega \frac{\partial n}{\partial \mu}\bigg|_{\lambda_1} = -\frac{\partial^2\Gamma}{\partial \mu^2}\bigg|_{\chi_{1o}} - \frac{\partial^2\Gamma}{\partial \mu \partial \chi_{1o}}\bigg| \frac{\partial \chi_{1o}}{\partial \mu}\bigg|_{\lambda_1} \tag{10.73}$$

and we use the following identity (with $\partial(x, y)/\partial(\alpha, \beta)$ we indicate the Jacobian):

$$\frac{\partial \chi_{1o}}{\partial \mu}\bigg|_{\lambda_1} = \frac{\partial(\chi_{1o}, \lambda_1)}{\partial(\mu, \lambda_1)} = \frac{\partial(\chi_{1o}, \lambda_1)}{\partial(\chi_{1o}, \mu)} \frac{\partial(\chi_{1o}, \mu)}{\partial(\mu, \lambda_1)} = -\frac{\partial \lambda_1}{\partial \mu}\bigg|_{\chi_{1o}} \frac{\partial \chi_{1o}}{\partial \lambda_1}\bigg|_{\mu} \tag{10.74}$$

and

$$\frac{\partial \lambda_1}{\partial \mu}\bigg|_{\chi_{1o}} = \frac{\partial^2\Gamma}{\partial \chi_{1o}\partial \mu}, \tag{10.75}$$

while by differentiating (10.71) with respect to χ_{1o} at constant μ we have:

$$\frac{\partial^2\Gamma}{\partial \chi_{1o}^2}\bigg|_{\mu} \left(\frac{\partial^2\Gamma}{\partial \chi_{1o}^2}\bigg|_{\mu}\right)^{-1} = 1 \quad \Rightarrow \quad \frac{\partial \chi_{1o}}{\partial \lambda_1}\bigg|_{\mu} = \left(\frac{\partial^2\Gamma}{\partial \chi_{1o}^2}\bigg|_{\mu}\right)^{-1} \tag{10.76}$$

Putting all pieces together we have:

$$\beta\Omega \frac{\partial n}{\partial \mu}\bigg|_{\lambda_1} = \frac{\left(\frac{\partial^2\Gamma}{\partial \chi_{1o}\partial \mu}\right)^2 - \frac{\partial^2\Gamma}{\partial \mu^2}\big|_{\chi_{1o}} \frac{\partial^2\Gamma}{\partial \chi_{1o}^2}\big|_{\mu}}{\frac{\partial^2\Gamma}{\partial \chi_{1o}^2}\big|_{\mu}} \tag{10.77}$$

1034   Comparison of the RG c and the Macroscopic c

In our formalism we have different values of the sound velocity at different steps of the RG procedure. The sound velocity entering (9.133) and (9.134) is defined in terms of the running couplings by the following expression:

$$c^2 = \frac{v_{11} z_{22}}{v_{11} u_{22} + w_{12}^2}.$$

(10.78)

We have found constraints on the limiting $k \to 0$ value of the running couplings, so we can try to substitute the limiting expressions for them in the equation for c. The asymptotic values for all these couplings is fixed by the Ward identities:

$$\begin{cases} v_{11} & \to & \dfrac{1}{\beta\Omega} \dfrac{\partial^2\Gamma}{\partial\chi_{1o}^2}\Big|_\mu \\[2mm] w_{12} & \to & \dfrac{1}{\beta\Omega\chi_{1o}} \dfrac{\partial^2\Gamma}{\partial\chi_{1o}\partial\mu} \\[2mm] u_{22} & \to & -\dfrac{1}{\beta\Omega\chi_{1o}^2} \dfrac{\partial^2\Gamma}{\partial\mu^2}\Big|_{\chi_{1o}} \\[2mm] z_{22} & \to & \dfrac{1}{\beta\Omega} \dfrac{2n_s}{\chi_{1o}^2}. \end{cases}$$

(10.79)

By substituting (10.79) into (10.78) we obtain the correct macroscopic sound velocity for the Bose gas (cf. (10.70) and (10.77)).

We report also an interesting identity:

$$\frac{v_{11}}{w_{12}} = \frac{\frac{\partial^2\Gamma}{\partial\chi_{1o}^2}\Big|_\mu}{\frac{1}{\chi_{1o}}\frac{\partial^2\Gamma}{\partial\chi_{1o}\partial\mu}} = -\chi_{1o}\frac{\partial\mu}{\partial\chi_{1o}}\Big|_\lambda,$$

(10.80)

and by the definition of the condensate density $n_0 = \chi_{1o}^2$ we obtain:

$$\frac{v_{11}}{w_{12}} = -2n_0 \frac{d\mu}{dn_0}\Big|_\lambda$$

(10.81)

which defines the condensate compressibility $(dn_0/d\mu)_\lambda$.

In the following we will study the flow of the running coupling present in (10.79) and we will see that the asymptotic physical limit for the sound velocity is recovered with a non trivial scaling of the running coupling. This means that it is just the sum of the singular terms for the vertex functions that determines these quantities.

Chapter 11

Renormalization Group Treatment

11.1 Renormalization Group Equations

Many different papers were devoted to study critical behavior of the Bose system with the RG approach [103–105], on the other hand the study of the IR divergence in the classical system of spin for vanishing external magnetic field is also a well studied subject [106–111]. By contrast the problem on the IR divergences for the Bose gas in the broken symmetry phase has long delayed for the reasons explained in the Introduction.

In this Section we calculate explicitly the one-loop RG equation with dimensional regularization. We perform an ϵ-expansion and apply the minimal subtraction technique. We then solve the one-loop RG equations and find the IR behavior of the system for $d > 1$ [112]. The one-loop solution suggests a way in which the theory can be generalized at all orders; to this purpose we then discuss the renormalized perturbation theory. The use of the Ward Identities turns out to be essential to give rigorous constraints on the scaling of the running couplings.

1111 Technical Points

We will calculate the one-loop diagrams within the dimensional regularization scheme. To this purpose we will need to integrate functions of the following form $(\beta \to \infty)$:

$$\frac{1}{\beta\Omega}\sum_q = \frac{1}{\beta\Omega}\Omega\frac{\beta}{2\pi}\int_{-\infty}^{+\infty} d\omega_\nu \int \frac{d^d\mathbf{q}}{(2\pi)^d} = c_o \int_{-\infty}^{\infty} \frac{dq_0}{2\pi} \int \frac{d^d\mathbf{q}}{(2\pi)^d} = c_o \int \frac{d^{d+1}q}{(2\pi)^{d+1}}.$$

$$(11.1)$$

To perform the integrations we need the following identities (Feynman trick) (see for instance Ref. [98]):

$$
\frac{1}{a_1^{\alpha_1} a_2^{\alpha_2} \dots a_n^{\alpha_n}} = \int_0^1 dx_1 \int_0^1 dx_2 \dots \int_0^1 dx_n
$$

$$
\times \frac{x_1^{\alpha_1-1} x_2^{\alpha_2-1} \dots x_n^{\alpha_n-1}}{[x_1 a_1 + x_2 a_2 + \dots + x_n a_n]^{\alpha_1+\alpha_2+\dots+\alpha_n}} \delta(x_1 + x_2 + \dots + x_n - 1)
$$

$$
\times \frac{\Gamma(\alpha_1 + \alpha_2 + \dots + \alpha_n)}{\Gamma(\alpha_1)\Gamma(\alpha_2)\dots\Gamma(\alpha_n)} \tag{11.2}
$$

$$
\int_0^{+\infty} \frac{x^{d-1} dx}{(1+x^2)^\alpha} = \frac{\Gamma(d/2)\Gamma(\alpha - d/2)}{2\Gamma(\alpha)} \tag{11.3}
$$

$$
\int dq F(q+k) = \int dq F(q) \tag{11.4}
$$

$$
\int dq F(\lambda q) = |\lambda|^{-d} \int dq F(q). \tag{11.5}
$$

By using (11.2)-(11.5) we obtain:

$$
\int \frac{d^d q}{(2\pi)^d} \frac{1}{(q^2 + 2kq + m^2)^\alpha} = K_d \frac{\Gamma(d/2)\Gamma(\alpha - d/2)}{2\Gamma(\alpha)} (m^2 - k^2)^{d/2-\alpha}. \tag{11.6}
$$

Here K_d is the surface of the unit sphere in d dimension divided by $(2\pi)^d$: $K_d = C_d/(2\pi)^d = 2/[(4\pi)^{d/2}\Gamma(d/2)]$ (in particular $K_4 = 1/(8\pi^2)$) and Γ in these equations is the Euler's Gamma function analytically continued in the complex plane:

$$
\Gamma(z) = \int_0^{+\infty} dt\, e^{-t} t^{z-1} \qquad \text{Re}(z) > 0. \tag{11.7}
$$

From Eq. (11.7) it is easy to verify that the simple relation holds:

$$
\Gamma(z+1) = z\Gamma(z). \tag{11.8}
$$

$\Gamma(1) = 1$ so that $\Gamma(n) = (n-1)!$. From (11.8) we obtain also that the Gamma function has poles when its argument is a negative integer or 0:

$$
\Gamma(z) = \frac{\Gamma(z+1)}{z} = \frac{\Gamma(z+2)}{z(z+1)} \tag{11.9}
$$

$$
\lim_{z\to 0} \Gamma(z) \approx \frac{1}{z} \tag{11.10}
$$

$$
\lim_{z\to -1} \Gamma(z) \approx -\frac{1}{z+1}. \tag{11.11}
$$

In the following we shall consider the integrals:

$$\int \frac{dq}{(2\pi)^d} \frac{F(k,q)}{q^2(k+q)^2} = \frac{\Gamma(2)}{\Gamma(1)\Gamma(1)} \int \frac{dq}{(2\pi)^d} \int_0^1 dx \frac{F(k,q)}{[x(k+q)^2 + (1-x)q^2]^2}$$

$$= \int_0^1 dx \int \frac{dq}{(2\pi)^d} \frac{F(k,q)}{[q^2 + 2kqx + k^2x]^2}$$

$$= \int_0^1 dx \int \frac{dq}{(2\pi)^d} \frac{F(k,q-xk)}{[q^2 + k^2x(1-x)]^2}$$

$$= \int_0^1 dx \int \frac{dq}{(2\pi)^d} \frac{F_s(k,q-xk)}{[q^2 + k^2x(1-x)]^2} \qquad (11.12)$$

where

$$F_s(k,q-xk) \equiv [F(k,q-kx) + F(k,-q-kx)]/2 \qquad (11.13)$$

So we define the following average:

$$\langle g(q,k,x) \rangle_f = \int_0^1 dx \int \frac{dq}{(2\pi)^d} \frac{f(x)g(k,q,x)}{[q^2 + k^2x(1-x)]^2} \qquad (11.14)$$

We calculate two important examples of these averages:

$$\langle 1 \rangle_f = \int_0^1 dx \int \frac{dq}{(2\pi)^{d+1}} \frac{f(x)}{[q^2 + k^2x(1-x)]^2}$$

$$= \frac{K_{d+1}}{2}\Gamma\left(\frac{d+1}{2}\right)\Gamma\left(2-\frac{d+1}{2}\right)\int_0^1 dx[k^2x(1-x)]^{\frac{d+1}{2}-2}f(x)$$

$$= (k^2)^{-\epsilon/2}K_{4-\epsilon}P(\epsilon)I_f(\epsilon) \qquad (11.15)$$

where we have defined $\epsilon = 4 - (d+1) = 3 - d$,

$$I_f(\epsilon) = \int_0^1 dx[x(1-x)]^{-\epsilon/2}f(x) \qquad (11.16)$$

and

$$P(\epsilon) = \frac{1}{2}\Gamma\left(2-\frac{\epsilon}{2}\right)\Gamma\left(\frac{\epsilon}{2}\right)$$

$$= \frac{1}{2}\left(1-\frac{\epsilon}{2}\right)\Gamma\left(1-\frac{\epsilon}{2}\right)\frac{2}{\epsilon}\Gamma\left(1+\frac{\epsilon}{2}\right)$$

$$= \frac{1}{\epsilon}(1-\epsilon/2)\Gamma(1-\epsilon/2)\Gamma(1+\epsilon/2). \qquad (11.17)$$

Using the results (11.15)-(11.17) we can easily write the result for the following average:

$$\langle q^2 \rangle_f = (k^2)^{\frac{2-\epsilon}{2}}K_{4-\epsilon}P(\epsilon-2)I_f(\epsilon-2) \qquad (11.18)$$

From the invariance under exchange of the various components of q_i we can also write the averages of q_0^2 and of $\mathbf{q}^2$:

$$\langle \mathbf{q}^2 \rangle_f = \frac{d}{d+1} \langle q^2 \rangle_f = \frac{3-\epsilon}{4-\epsilon}(k^2)^{\frac{2-\epsilon}{2}} K_{4-\epsilon} P(\epsilon-2) I_f(\epsilon-2) \quad (11.19)$$

$$\langle q_0^2 \rangle_f = \frac{1}{d+1} \langle q^2 \rangle_f = \frac{1}{4-\epsilon}(k^2)^{\frac{2-\epsilon}{2}} K_{4-\epsilon} P(\epsilon-2) I_f(\epsilon-2) \quad (11.20)$$

We consider the lowest order in ϵ of the functions $P(\epsilon)$, $P(\epsilon-2)$, $I_f(\epsilon)$, and $I_f(\epsilon-2)$:

$$P(\epsilon) = \frac{1}{\epsilon}[1 - \epsilon/2 + O(\epsilon^2)] \quad (11.21)$$

$$P(\epsilon-2) = -\frac{1}{\epsilon}(2-\epsilon/2)\Gamma(1-\epsilon/2)\Gamma(1+\epsilon/2) = -\frac{2}{\epsilon}\left[1 - \frac{\epsilon}{4} + O(\epsilon^2)\right] \quad (11.22)$$

$$I_f(\epsilon) = \int_0^1 dx\, f(x) - \frac{\epsilon}{2}\int_0^1 dx\, \ln[x(1-x)]f(x) + O(\epsilon^2) \quad (11.23)$$

$$I_f(\epsilon-2) = \int_0^1 dx\, f(x)x(1-x) - \frac{\epsilon}{2}\int_0^1 dx\, x(1-x)\ln[x(1-x)]f(x) + O(\epsilon^2) \quad (11.24)$$

1112   Greens Functions After Rescaling by c_o

We recall that we must "eliminate" c_o from the Green's functions. This scaling gives the following expressions for the $\mathcal{G}_{ij}$ (in the asymptotic $k \to 0$ region):

$$\mathcal{G}_{11}(k_0, \mathbf{k}) = \frac{u_{22}c_o^2 k_0^2 + z_{22}\mathbf{k}^2}{v_{11}z_{22}}\frac{1}{k^2} \quad (11.25)$$

$$\mathcal{G}_{12}(k_0, \mathbf{k}) = -\frac{c_o w_{12}}{v_{11}z_{22}}\frac{k_0}{k^2} \quad (11.26)$$

$$\mathcal{G}_{22}(k_0, \mathbf{k}) = \frac{1}{z_{22}}\frac{1}{k^2} \quad (11.27)$$

[cf. (9.133)-(9.134)] with $k^2 = \mathbf{k}^2 + c_0^2 k_0^2/c^2$

1113   Diagrams Calculation

We are left with the calculation of the diagrams depicted in Fig. 11.1 for the non-composite quadratic vertex parts [cf. (9.68)].

$$\begin{aligned}
A_{11}(k) &= \frac{1}{2}\frac{1}{\beta\Omega}\sum_q \mathcal{G}_{22}(q)\mathcal{G}_{22}(k+q)v_{122}^2 \\
&= \frac{1}{2}\frac{v_{11}^2}{\chi_{1o}^2}c_o\frac{1}{z_{22}^2}\int \frac{d^{d+1}q}{(2\pi)^{d+1}}\frac{1}{q^2(k+q)^2} \\
&= \frac{c_o v_{11}^2}{2\chi_{1o}^2 z_{22}^2}g_{11}(k) \quad (11.28)
\end{aligned}$$

where from (11.12)-(11.15) we obtain:

$$g_{11}(k) = \int \frac{d^{d+1}q}{(2\pi)^{d+1}} \frac{1}{q^2(k+q)^2} = \langle 1 \rangle_1 = (k^2)^{-\epsilon/2} K_{4-\epsilon} P(\epsilon) I_1(\epsilon). \quad (11.29)$$

$$\begin{aligned}
A_{12}(k) &= \frac{1}{\beta\Omega} \sum_q \mathcal{G}_{12}(-q)\mathcal{G}_{22}(k-q)v_{122}^2 \\
&= -\frac{v_{11}^2}{\chi_{1o}^2} c_o \frac{c_o w_{12}}{v_{11}z_{22}^2} \int \frac{d^{d+1}q}{(2\pi)^{d+1}} \frac{q_0}{q^2(k+q)^2} \\
&= -\frac{w_{12}v_{11}c_o^2}{\chi_{1o}^2 z_{22}^2} g_{12}(k), \quad (11.30)
\end{aligned}$$

where

$$\begin{aligned}
g_{12}(k) &= \int \frac{d^{d+1}q}{(2\pi)^{d+1}} \frac{q_0}{q^2(k+q)^2} = \langle (q_0 - k_0 x)_s \rangle_1 \\
&= -k_0 \langle 1 \rangle_x = -k_0 (k^2)^{-\epsilon/2} K_{4-\epsilon} P(\epsilon) I_x(\epsilon). \quad (11.31)
\end{aligned}$$

$$\begin{aligned}
A_{22}(k) &= \frac{1}{\beta\Omega} \sum_q \mathcal{G}_{11}(q)\mathcal{G}_{22}(k-q)v_{122}^2 \\
&= \frac{v_{11}^2}{\chi_{1o}^2} c_o \frac{1}{v_{11}z_{22}^2} \int \frac{d^{d+1}q}{(2\pi)^{d+1}} \frac{u_{22}c_o^2 q_0^2 + z_{22}\mathbf{q}^2}{q^2(k+q)^2} \\
&= \frac{cv_{11}}{\chi_{1o}^2 z_{22}^2} [u_{22}c_o^2 g_{22}(k) + z_{22}h_{22}(k)], \quad (11.32)
\end{aligned}$$

where

$$\begin{aligned}
g_{22}(k) &= \int \frac{d^{d+1}q}{(2\pi)^{d+1}} \frac{q_0^2}{q^2(k+q)^2} = \langle (q_0 - k_0 x)_s^2 \rangle_1 \\
&= k_0^2 \langle 1 \rangle_{x^2} + \langle q_0^2 \rangle_1 \\
&= k_0^2 (k^2)^{-\epsilon/2} K_{4-\epsilon} P(\epsilon) I_{x^2}(\epsilon) + \frac{1}{4-\epsilon} (k^2)^{-\epsilon/2+1} K_{4-\epsilon} P(\epsilon-2) I_1(\epsilon-2)
\end{aligned}$$

$$(11.33)$$

and

$$\begin{aligned}
h_{22}(k) &= \int \frac{d^{d+1}q}{(2\pi)^{d+1}} \frac{\mathbf{q}^2}{q^2(k+q)^2} = \langle (\mathbf{q} - \mathbf{k}x)_s^2 \rangle_1 \\
&= \mathbf{k}^2 \langle 1 \rangle_{x^2} + \langle \mathbf{q}^2 \rangle_1 \\
&= \mathbf{k}^2 (k^2)^{-\epsilon/2} K_{4-\epsilon} P(\epsilon) I_{x^2}(\epsilon) + \frac{3-\epsilon}{4-\epsilon} (k^2)^{-\epsilon/2+1} K_{4-\epsilon} P(\epsilon-2) I_1(\epsilon-2).
\end{aligned}$$

$$(11.34)$$

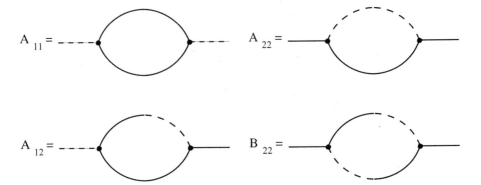

Figure 11.1: One-loop diagrams for the Γ_{11}, Γ_{12}, and Γ_{22} considered in the text.

$$B_{22}(k) = \frac{1}{\beta\Omega} \sum_q \mathcal{G}_{12}(q)\mathcal{G}_{21}(k-q)v_{122}^2$$

$$= \frac{v_{11}^2}{\chi_{1o}^2} c_o \left(\frac{c_o w_{12}}{v_{11} z_{22}}\right)^2 \int \frac{d^{d+1}}{(2\pi)^{d+1}} \frac{q_0(k_0 + q_0)}{q^2(k+q)^2}$$

$$= \frac{c_o^3 w_{12}^2}{\chi_{1o}^2 z_{22}^2} j_{22}(k), \tag{11.35}$$

where

$$j_{22}(k) = \int \frac{d^{d+1}q}{(2\pi)^{d+1}} \frac{q_0(q_0 + k_0)}{q^2(k+q)^2} = \langle[(q_0 - k_0 x)(q_0 - k_0 x + k_0)]_s\rangle_1$$

$$= \langle[q_0^2 - k_0(x-1)q_0 - k_0 x q_0 + k_0^2 x(x-1)]_s\rangle_1 = \langle q_0^2 + k_0^2 x(x-1)\rangle_1$$

$$= \langle q_0^2\rangle_1 + k_0^2 \langle 1\rangle_{x(x-1)}$$

$$= k_0^2 (k^2)^{-\epsilon/2} K_{4-\epsilon} P(\epsilon) I_{x(x-1)}(\epsilon)$$

$$+ \frac{1}{4-\epsilon}(k^2)^{-\epsilon/2+1} K_{4-\epsilon} P(\epsilon-2) I_1(\epsilon-2). \tag{11.36}$$

The vertex functions at the order of one loop are thus given by the following expressions [cf. (9.120) and (9.52)]:

$$\bar{\Gamma}_{11}(k) = v_{11} - \frac{v_{11}^2 c_o}{2\chi_{1o}^2 z_{22}^2} g_{11}(k) \tag{11.37}$$

$$\bar{\Gamma}_{12}(k) = w_{12} c_o k_0 + \frac{w_{12} v_{11} c_o^2}{\chi_{1o}^2 z_{22}^2} g_{12}(k) \tag{11.38}$$

$$\bar{\Gamma}_{22}(k) = u_{22} c_o^2 k_0^2 + z_{22} \mathbf{k}^2 - \frac{v_{11} c_o^3 u_{22}}{\chi_{1o}^2 z_{22}^2} g_{22}(k) - \frac{v_{11} c_o z_{22}}{\chi_{1o}^2 z_{22}^2} h_{22}(k) - \frac{c_o^3 w_{12}^2}{\chi_{1o}^2 z_{22}^2} j_{22}(k)$$

$$\tag{11.39}$$

1114 Dimensionless Running Couplings Denition

To proceed further we must eliminate the constant c_o from (11.37)-(11.39). This is done simply by recalling the rescaling of the fields performed in section 9.1.3. From (9.20) and (9.46) we find the powers of c_o we have to introduce in each running coupling.

$$[v_n] = -nd/2 + d + 2, \tag{11.40}$$

and

$$\left[\bar{\Gamma}^{(n_1+n_2)}_{1...12...2}\right] = -n_1(d+1)/2 - n_2(d-1)/2 + d + 1, \tag{11.41}$$

so the change in the dimension of the running couplings is:

$$\left[\tilde{v}^{n_1+n_2}_{1...12...2}\right] - [v_{n_1+n_2}] = -n_1/2 + n_2/2 - 1. \tag{11.42}$$

We indicate with a tilde the new running couplings ($[c_o] = [k] = 1$):

$$\begin{cases}
v_{11} &= c_o^2 \tilde{v}_{11} \\
v_{122} &= c_o^{1/2} \tilde{v}_{122} \\
v_{2222} &= c_o^{-1} \tilde{v}_{2222} \\
\chi_{1o} &= c_o^{3/2} \tilde{\chi}_{1o} \\
v_{12} &= c_o \tilde{v}_{12} \\
w_{12} &= \tilde{w}_{12} \\
u_{22} &= \tilde{u}_{22}/c_o^2 \\
z_{22} &= \tilde{z}_{22}.
\end{cases} \tag{11.43}$$

Where the factors for z_{22}, u_{22} and w_{12} comes from the following identity:

$$\bar{\Gamma}_{12} = v_{12} + w_{12}\omega = v_{12} + w_{12}c_o k_0, \tag{11.44}$$

and the variation in the powers of c_o of $\bar{\Gamma}_{12}$ is -1. In the same way:

$$\bar{\Gamma}_{22} = v_{22} + u_{22}c_o^2 k_0^2 + z_{22}\mathbf{k}^2. \tag{11.45}$$

With these results we can write the one-loop corrections to the $\tilde{\Gamma}$:

$$\begin{cases}
\tilde{\Gamma}_{11}(k) &= \tilde{v}_{11} - \dfrac{\tilde{v}_{11}^2}{2\tilde{\chi}_{1o}^2 \tilde{z}_{22}^2} g_{11}(k) \\[2ex]
\tilde{\Gamma}_{12}(k) &= \tilde{w}_{12}k_0 + \dfrac{\tilde{w}_{12}\tilde{v}_{11}}{\tilde{\chi}_{1o}^2 \tilde{z}_{22}^2} g_{12}(k) \\[2ex]
\tilde{\Gamma}_{22}(k) &= \tilde{u}_{22}k_0^2 + \tilde{z}_{22}\mathbf{k}^2 - \dfrac{\tilde{v}_{11}\tilde{u}_{22}}{\tilde{\chi}_{1o}^2 \tilde{z}_{22}^2} g_{22}(k) - \dfrac{\tilde{v}_{11}\tilde{z}_{22}}{\tilde{\chi}_{1o}^2 \tilde{z}_{22}^2} h_{22}(k) - \dfrac{\tilde{w}_{12}^2}{\tilde{\chi}_{1o}^2 \tilde{z}_{22}^2} j_{22}(k)
\end{cases} \tag{11.46}$$

Note that the running couplings are not dimensionless for $d \neq 3$:

$$\begin{aligned}
[\tilde{v}_{122}] &= \epsilon/2 & [\tilde{v}_{11}] &= [\tilde{z}_{22}] = [\tilde{w}_{12}] = [\tilde{u}_{22}] = 0 \\
[\tilde{v}_{2222}] &= \epsilon & [\tilde{\chi}_{1o}] &= -\epsilon/2
\end{aligned} \tag{11.47}$$

To understand better the solution of the RG equations we can write the one-loop correction to $\tilde{v}_{122}$ and $\tilde{v}_{2222}$. This is possible in a simple way by using the Ward identity (8.75):

$$\bar{\Gamma}_{221}(0,k)\chi_{1o} - \bar{\Gamma}_{11}(k) + \bar{\Gamma}_{22}(-k) = 0 \qquad (11.48)$$

$\bar{\Gamma}_{22}(0) = 0$ and the difference between $\Gamma(k,k)$ and $\Gamma(k,0)$ should be of order $k^2 \ln k$ (in $d = 3$):

$$\tilde{\Gamma}_{221}(0,k) = \frac{\tilde{\Gamma}_{11}(k)}{\chi_{1o}} = \tilde{v}_{122} - \frac{\tilde{v}_{122}^2}{2\chi_{1o}\tilde{z}_{22}^2}g_{11}(k). \qquad (11.49)$$

We can obtain the same result for Γ_{2222}:

$$\tilde{\Gamma}_{2222}(0,k_2,k_3) = 3\frac{\tilde{\Gamma}_{122}(k_2,k_3)}{\chi_{1o}} = 3\tilde{v}_{2222} - \frac{\tilde{v}_{2222}^2}{6\tilde{z}_{22}^2}g_{11}(k_2) \qquad (11.50)$$

We now substitute the lowest order (in ϵ) expressions for g_{11}, g_{12}, g_{22}, h_{22} and j_{22}. Note that the factor $K_{4-\epsilon}$ can be absorbed in the running couplings to simplify the expressions:

$$\begin{cases} \tilde{v}_{11} & \rightarrow & \tilde{v}_{11}K_{4-\epsilon} \\ \tilde{w}_{12} & \rightarrow & \tilde{w}_{12}(K_{4-\epsilon})^{1/2} \end{cases} \qquad (11.51)$$

$$\begin{cases} g_{11}(k) & = & \dfrac{\kappa^{-\epsilon}}{\epsilon}[1 + O(\epsilon)]\,I_1(0)K_{4-\epsilon} \\[2mm] g_{12}(k) & = & -k_0\dfrac{\kappa^{-\epsilon}}{\epsilon}[1 + O(\epsilon)]\,K_{4-\epsilon}I_x(0) \\[2mm] g_{22}(k) & = & k_0^2\dfrac{\kappa^{-\epsilon}}{\epsilon}K_{4-\epsilon}\,[I_{x^2}(0) - I_1(-2)/2] - \mathbf{k}^2\dfrac{\kappa^{-\epsilon}}{\epsilon}K_{4-\epsilon}I_1(-2)/2 \\[2mm] h_{22}(k) & = & \mathbf{k}^2\dfrac{\kappa^{-\epsilon}}{\epsilon}K_{4-\epsilon}\,[I_{x^2}(0) - 3I_1(-2)/2] - 3k_0^2\dfrac{\kappa^{-\epsilon}}{\epsilon}K_{4-\epsilon}I_1(-2)/2 \\[2mm] j_{22}(k) & = & k_0^2\dfrac{\kappa^{-\epsilon}}{\epsilon}K_{4-\epsilon}\,[I_{x^2}(0) - I_x(0) - I_1(-2)/2] - \mathbf{k}^2\dfrac{\kappa^{-\epsilon}}{\epsilon}K_{4-\epsilon}I_1(-2)/2\,, \end{cases} \qquad (11.52)$$

where we have defined $\kappa = \sqrt{k^2}$. The following table gives the values for I:

$$\begin{cases} I_1(0) & = & 1 \\ I_x(0) & = & 1/2 \\ I_{x^2}(0) & = & 1/3 \\ I_1(-2) & = & 1/6\,. \end{cases} \qquad (11.53)$$

In this way we can calculate the one-loop corrections to the Γ's. In particular for Γ_{22} we obtain:

$$\begin{aligned} \tilde{\Gamma}_{22}^{(1)}(k) & = & \frac{\mathbf{k}^2\kappa^{-\epsilon}}{12\tilde{z}_{22}^2\tilde{\chi}_{1o}^2\epsilon}\left(\tilde{v}_{11}\tilde{u}_{22} + \tilde{w}_{12}^2 - \tilde{v}_{11}\tilde{z}_{22}\right) \\[2mm] & & -\frac{k_0^2\kappa^{-\epsilon}}{2\tilde{z}_{22}^2\tilde{\chi}_{1o}^2\epsilon}\left[\left(\tilde{v}_{11}\tilde{u}_{22} + \tilde{w}_{12}^2 - \tilde{v}_{11}\tilde{z}_{22}\right)/2 - \tilde{w}_{12}^2\right], \qquad (11.54) \end{aligned}$$

but now recall that the sound velocity is given by [cf. (10.78)]

$$1 = \frac{\tilde{v}_{11}\tilde{z}_{22}}{\tilde{v}_{11}\tilde{u}_{22} + \tilde{w}_{12}^2} \Rightarrow \tilde{v}_{11}\tilde{u}_{22} + \tilde{w}_{12}^2 - \tilde{v}_{11}\tilde{z}_{22} = 0 \,, \tag{11.55}$$

so Eq.(11.54) simplifies greatly. The final result for the vertex functions is the following:

$$\begin{cases} \tilde{\Gamma}_{11}(k) &= \tilde{v}_{11} - \dfrac{\tilde{v}_{11}^2 \kappa^{-\epsilon}}{2\tilde{\chi}_{1o}^2 \tilde{z}_{22}^2 \epsilon} \\[2mm] \tilde{\Gamma}_{12}(k) &= \tilde{w}_{12}k_0 - \dfrac{\tilde{w}_{12}\tilde{v}_{11}\kappa^{-\epsilon}}{2\tilde{\chi}_{1o}^2 \tilde{z}_{22}^2 \epsilon}k_0 \\[2mm] \tilde{\Gamma}_{22}(k) &= \tilde{u}_{22}k_0^2 + \tilde{z}_{22}\mathbf{k}^2 + \dfrac{\tilde{w}_{12}^2 \kappa^{-\epsilon}}{2\tilde{\chi}_{1o}^2 \tilde{z}_{22}^2 \epsilon}k_0^2 \end{cases} \tag{11.56}$$

A final remark is necessary. If we leave the possibility that the renormalized c will be different from the bare c_o we must introduce in (11.56) a c/c_o term for each loop of integration (in this case one). This comes from the integral in $d\omega = cdk_0$ that must be performed at each loop. We will show later that c does not scale, so that all these terms can be neglected throughout.

1115   RG Approach

In this Section we will briefly review the RG approach [113, 98] that we will use. We define the set of running coupling $\{\lambda_i\}$, and we consider the case in which the theory is regularized in the UV with a cutoff Λ (then we will eliminate it in favor of the dimensional regularization). We suppose to have a set of normalization conditions, for example:

$$\tilde{\Gamma}_{11}(k)|_{k^2=\kappa^2} = g_1 \,. \tag{11.57}$$

We do not consider the possibility of wave function renormalization. We define the renormalized vertex part in the following way:

$$\tilde{\Gamma}_R(k_i, g_i(\lambda_i, \kappa, \Lambda), \kappa) = \tilde{\Gamma}(k_i, \lambda_i, \Lambda) \,. \tag{11.58}$$

From Eq.(11.58) we can obtain the RG equations by differentiating with respect to κ at λ_i constant:

$$\kappa \left.\frac{\partial}{\partial\kappa}\right|_\lambda \tilde{\Gamma}(k_i, \lambda, \kappa) = 0 = \kappa \left.\frac{\partial}{\partial\kappa}\right|_\lambda \tilde{\Gamma}_R(k_i, g, \kappa)$$

$$= \left[\kappa \left.\frac{\partial}{\partial\kappa}\right|_g + \kappa \left.\frac{\partial g_i}{\partial\kappa}\right|_\lambda \frac{\partial}{\partial g_i} \right] \tilde{\Gamma}_R(k_i, g, \kappa) \,. \tag{11.59}$$

If the dimension of the coupling g_i is α_i, we can introduce the dimensionless couplings u_i and u_i^o:

$$g_i = \kappa^{\alpha_i} u_i \qquad \lambda_i = \kappa^{\alpha_i} u_i^o \,. \tag{11.60}$$

In this way we obtain:

$$\kappa \left. \frac{\partial g_i}{\partial \kappa} \right|_\lambda = \alpha_i g_i + \kappa^{\alpha_i} \kappa \left. \frac{\partial u_i}{\partial \kappa} \right|_\lambda , \qquad (11.61)$$

and

$$\kappa \left. \frac{\partial}{\partial \kappa} \right|_g = \kappa \left. \frac{\partial}{\partial \kappa} \right|_u - \kappa \left. \frac{\partial g_i}{\partial \kappa} \right|_u \left. \frac{\partial}{\partial g_i} \right|_\kappa . \qquad (11.62)$$

So if we define the beta function in the following way:

$$\beta_i(u) \equiv \kappa \left. \frac{\partial u_i}{\partial \kappa} \right|_\lambda , \qquad (11.63)$$

we obtain the RG equation in its standard form:

$$\left[\kappa \left. \frac{\partial}{\partial \kappa} \right|_u + \beta_i(u) \left. \frac{\partial}{\partial u_i} \right|_\kappa \right] \Gamma_R(k_i, u_i, \kappa) = 0 . \qquad (11.64)$$

We can verify that the following equation holds:

$$\Gamma_R(k_i, u, \kappa) = \Gamma_R(k_i, u(\rho), \kappa\rho) , \qquad (11.65)$$

where

$$\rho \frac{du_i(\rho)}{d\rho} = \beta_i(u) . \qquad (11.66)$$

In fact for $\rho = 1$ it is trivially satisfied, and by differentiating with respect to ρ we obtain the RG equation (11.64). Note that (11.65) can also be written in the following form:

$$\Gamma_R(\rho k_i, u, \kappa) = \Gamma_R(\rho k_i, u(\rho), \kappa\rho) = \rho^\alpha \Gamma_R(k_i, u(\rho), \kappa) , \qquad (11.67)$$

where α is the dimension of the vertex function. So to study the IR behavior of the vertex functions we simply have to scale it with its bare dimension α and substitute the new value of the running couplings $u(\rho)$. In particular in the case that the renormalization condition is

$$\left. \tilde{\Gamma}_R(k, \kappa) \right|_{k=\kappa} = g_i = \kappa^{\alpha_i} u_i \qquad (11.68)$$

we obtain

$$\tilde{\Gamma}_R(\rho\kappa, u, \kappa) = \kappa^{\alpha_i} \rho^{\alpha_i} u_i(\rho) , \qquad (11.69)$$

so that, if near the fixed point the running coupling is flowing to zero with an exponent y_i

$$u_i(\rho) \sim \rho^{y_i}, \qquad (11.70)$$

we get the real final scaling of the vertex part:

$$\tilde{\Gamma}_R(k) \sim k^{\alpha_i + y_i} . \qquad (11.71)$$

Note that if $y_i = 0$ the scaling of the vertex part is given by the bare dimension, while if y_i is exactly equal to $-\alpha_i$ the vertex part does not depend on k and is constant in the IR.

1116   Minimal Subtraction RG Equation

The normalization condition can be written in the following way:

$$\begin{cases} \tilde{\Gamma}_{11}(k)|_{k^2=\kappa^2} = v'_{11} \\ \dfrac{\tilde{\Gamma}_{12}(k)}{k_0}\bigg|_{k^2=\kappa^2} = w'_{12} \\ \dfrac{\tilde{\Gamma}_{22}(k)}{k_0^2}\bigg|_{k^2=\kappa^2} = u'_{22} \\ \dfrac{\tilde{\Gamma}_{22}(k)}{\mathbf{k}^2}\bigg|_{k^2=\kappa^2} = z'_{22} \end{cases} \tag{11.72}$$

In this way we have (at the order of one loop):

$$\begin{cases} v'_{11} = \tilde{v}_{11} - \dfrac{\tilde{v}_{11}^2 \kappa^{-\epsilon}}{2\tilde{\chi}_{1o}^2 \tilde{z}_{22}^2 \epsilon} \\[2ex] w'_{12} = \tilde{w}_{12} - \dfrac{\tilde{w}_{12}\tilde{v}_{11}\kappa^{-\epsilon}}{2\tilde{\chi}_{1o}^2 \tilde{z}_{22}^2 \epsilon} \\[2ex] u'_{22} = \tilde{u}_{22} + \dfrac{\tilde{w}_{12}^2 \kappa^{-\epsilon}}{2\tilde{\chi}_{1o}^2 \tilde{z}_{22}^2 \epsilon} \\[1ex] z'_{22} = \tilde{z}_{22}. \end{cases} \tag{11.73}$$

We must introduce a new coupling with a trivial scaling (by definition since we are working at fixed condensate density)

$$\chi'_{1o} = \tilde{\chi}_{1o} \tag{11.74}$$

We introduce now the dimensionless running couplings

$$\begin{cases} \bar{v}_{11} = v'_{11} \\ \bar{w}_{12} = w'_{12} \\ \bar{u}_{22} = u'_{22} \\ \bar{z}_{22} = z'_{22} \\ \bar{\chi}_{1o} = \chi'_{1o}\kappa^{\epsilon/2} \end{cases} \tag{11.75}$$

We apply the definition (11.63) and calculate the beta function of the running couplings:

$$
\begin{cases}
\rho \dfrac{d\bar{v}_{11}}{d\rho} = \beta_{v11} = \dfrac{\bar{v}_{11}^2}{2\bar{\chi}_{1o}^2 \bar{z}_{22}^2} \\[2ex]
\rho \dfrac{d\bar{w}_{12}}{d\rho} = \beta_{w12} = \dfrac{\bar{v}_{11}\bar{w}_{12}}{2\bar{\chi}_{1o}^2 \bar{z}_{22}^2} \\[2ex]
\rho \dfrac{d\bar{u}_{22}}{d\rho} = \beta_{u22} = -\dfrac{\bar{w}_{12}^2}{2\bar{\chi}_{1o}^2 \bar{z}_{22}^2} \\[2ex]
\rho \dfrac{d\bar{z}_{22}}{d\rho} = \beta_{z22} = 0 \\[2ex]
\rho \dfrac{d\bar{\chi}_{1o}}{d\rho} = \beta_{\chi_{1o}} = (\epsilon/2)\bar{\chi}_{1o}
\end{cases}
\tag{11.76}
$$

1117   Solution of the RG Equations

The system of equations can be solved easily. In particular, we can find a simple relation among the couplings:

$$
\bar{w}_{12}(\rho) = \frac{\bar{w}_{12}(1)}{\bar{v}_{11}(1)} \bar{v}_{11}(\rho) \tag{11.77}
$$

$$
\bar{u}_{22}(\rho) = \bar{u}_{22}(1) + \frac{\bar{w}_{12}(1)^2}{\bar{v}_{11}(1)} - \frac{\bar{w}_{12}(1)^2}{\bar{v}_{11}(1)^2} \bar{v}_{11}(\rho) \tag{11.78}
$$

$$
\bar{z}_{22}(\rho) = \bar{z}_{22}(1) \tag{11.79}
$$

These results are valid for any behavior of $\bar{v}_{11}(\rho)$ and $\bar{\chi}_{1o}(\rho)$.

We must now distinguish the case $\epsilon = 0$ and $\epsilon > 0$. For $\epsilon = 0$ (*i.e.* $d = 3$) we have:

$$
\bar{v}_{11}(\rho) = \frac{\bar{v}_{11}(1)}{1 - \frac{\bar{v}_{11}(1)}{2\bar{\chi}_{1o}(1)^2 \bar{z}_{22}(1)^2} \ln\rho} \sim -2 \frac{\bar{\chi}_{1o}(1)^2 \bar{z}_{22}(1)^2}{\ln \rho} \tag{11.80}
$$

$$
\bar{\chi}_{1o}(\rho) = \bar{\chi}_{1o}(1), \tag{11.81}
$$

while for $\epsilon > 0$ the result is:

$$
\bar{v}_{11}(\rho) = \frac{\bar{v}_{11}(1)}{1 + \frac{\bar{v}_{11}(1)}{2\bar{\chi}_{1o}(1)^2 \bar{z}_{22}(1)^2 \epsilon}(\rho^{-\epsilon} - 1)} \sim 2\bar{\chi}_{1o}(1)^2 \bar{z}_{22}(1)^2 \epsilon\rho^\epsilon \tag{11.82}
$$

$$
\bar{\chi}_{1o}(\rho) = \bar{\chi}_{1o}(1)\rho^{\epsilon/2}. \tag{11.83}
$$

In any case we have that the two couplings $\bar{v}_{11}$ and $\bar{w}_{12}$ vanish like $1/\ln \rho$ or as ρ^ϵ [114]. The other two couplings flow toward a finite value. ($\bar{z}_{22}$ remains constant.)

1118   Generalization at All Orders of the Constraints

Although we have derived (11.75)-(11.79) at the one-loop order, we *expect* them to hold exactly on physical ground owing to the identification (10.79) of the renormalization parameters with physical quantities [115]. To begin with, the $\rho \to 0$ value of $\bar{z}_{22}$ is the ratio n_s/n_0 of *finite* physical quantities so that divergences compensate each other in its expression leading to (11.79). Regarding (11.75), we recognize from (10.81) that the ratio $\bar{v}_{11}/\bar{w}_{12}$ reduces to $-2n_0/(dn_0/d\mu)_\lambda$ in the limit $k \to 0$. Here $(dn_0/d\mu)_\lambda$ is the "condensate compressibility" which has to be finite for stability. Concerning, finally, (11.76) we obtain from (10.79) and (10.78) that $c(\rho)^2$ reduces to c^2 in the limit $\rho \to 0$, where now $c^2 = 2n_s/(dn/d\mu)_\lambda$ is the square of the macroscopic sound velocity (n being the density and $n = n_s$ at zero temperature). By the very stability of the bosonic system, $c(\rho)$ is finite and does not scale with ρ, thus implying that $c(\rho) = c = c_0$ (apart from finite corrections originating from nonsingular terms that do not enter the RG flow). Exploiting further (11.75) and (11.79), we verify that (10.78) with $c(\rho)$ constant reduces to (11.76).

A *proof* of the above statements runs along the following lines.

$\bar{z}_{22}$ has to remain constant by inspection of the WI (10.54), which shows that the divergence of $\bar{z}_{22}$ expected by power counting is actually not present, since it is related to non diverging quantities. This point has been discussed after equation (10.54).

w_{12} can instead be identified with v_{11} via the WI (8.75) and (8.77), which relate w_{12} to $\Gamma_{22;0}$ and $\bar{v}_{11}$ to Γ_{122}, respectively. By inspection of the leading singular terms to all orders in perturbation theory, $\Gamma_{22;0}$ and Γ_{122} are then found to be proportional to each other. As a matter of fact the only way in which we can make an insertion of the longitudinal (type 1) leg is to use the bare interaction term $\bar{v}_{122}$ that connect the external longitudinal leg to two internal transverse legs. It is then clear that this corresponds to perform an insertion of density with the most relevant term, apart from the value of the bare external insertion that is different from the bare $\bar{v}_{122}$ (see Fig.11.2). We can write the chain of identities in the following way:

$$\chi_{1o}\Gamma_{11} \sim \Gamma_{122} \sim \frac{v_{122}}{\gamma_{22;0}}\Gamma_{22;0} \sim \frac{v_{122}}{\gamma_{22;0}}\frac{\partial\Gamma_{12}}{\partial\omega} \sim \frac{\chi_{1o}v_{11}}{\gamma_{22;0}}\frac{\partial\Gamma_{12}}{\partial\omega} \tag{11.84}$$

So we have:

$$\Gamma_{11} \sim \frac{v_{11}}{\gamma_{22;0}}\frac{\partial\Gamma_{12}}{\partial\omega} \tag{11.85}$$

In this way the logarithmic divergences that appear in the RG equation for $\bar{w}_{12}$ must also appear in the RG equation for $\bar{v}_{11}$, apart from a constant term. If the exact (all orders) RG equation for $\bar{v}_{11}$ is the following

$$\rho\frac{d\bar{v}_{11}}{d\rho} = f(\bar{v}_{11}, \bar{w}_{12}, \bar{u}_{22}) \tag{11.86}$$

then we can write the equation for $\bar{w}_{12}$:

$$\rho\frac{d\bar{w}_{12}}{d\rho} = Af(\bar{v}_{11}, \bar{w}_{12}, \bar{u}_{22})\frac{\bar{w}_{12}}{\bar{v}_{11}} \tag{11.87}$$

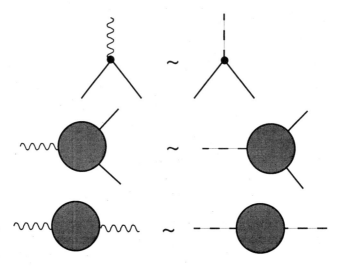

Figure 11.2: Asymptotic equivalence of $\Gamma_{22;0}$ with Γ_{122}, and $\Gamma_{;00}$ with Γ_{11}.

where A does not depend on ρ or on the other running couplings, and by making use of WI we substituted $\gamma_{22;0}$ with $\bar{w}_{12}$. From (11.86) and (11.87) we get:

$$\frac{d\bar{w}_{12}}{d\rho} / \frac{d\bar{v}_{11}}{d\rho} = A\frac{\bar{w}_{12}(\rho)}{\bar{v}_{11}(\rho)} \tag{11.88}$$

(11.88) implies that:

$$\frac{\bar{v}_{11}(\rho)}{\bar{w}_{12}(\rho)} = \text{constant}. \tag{11.89}$$

In a similar way, the invariance of $c(\rho)$ implied by (11.79) follows from the exact connection between the singular parts of $\Gamma_{;00}$ and Γ_{11}, associated respectively with u_{22} and v_{11}. The procedure leads to the following equation for $\bar{u}_{22}$:

$$\rho\frac{d\bar{u}_{22}}{d\rho} = Bf(\bar{v}_{11}, \bar{w}_{12}, \bar{u}_{22})\frac{\bar{w}_{12}^2}{\bar{v}_{11}^2} \tag{11.90}$$

using the result of (11.89) we obtain:

$$\bar{u}_{22}(\rho) = B'\bar{v}_{11}(\rho) + \text{constant} \tag{11.91}$$

So we are left eventually with only one independent running coupling: for instance $\bar{v}_{11}$.

The constraint found above for $\bar{v}_{11}$, $\bar{w}_{12}$, $\bar{z}_{22}$, and $\bar{u}_{22}$ fix completely the IR behavior of the two Green functions: $\mathcal{G}_{12}(k)$ and $\mathcal{G}_{22}(k)$. In fact in (11.26) and (11.27) there appears only combinations of running couplings that do not scale $(\bar{v}_{11}/\bar{w}_{12}$, $\bar{z}_{22}$, and $c^2 = \bar{v}_{11}\bar{z}_{22}/(\bar{v}_{11}\bar{u}_{22} + \bar{w}_{12}))$. So the resulting *exact* IR behavior is the following one:

$$\mathcal{G}_{12}(k) \sim \frac{\omega}{\omega^2 + c^2 \mathbf{k}^2} \tag{11.92}$$

$$\mathcal{G}_{22}(k) \sim \frac{1}{\omega^2 + c^2 \mathbf{k}^2} \tag{11.93}$$

$$\tag{11.94}$$

This behavior is valid at all orders of perturbation theory and does not depend on the behavior of $\tilde{v}_{11}$

1119   Validity of the Expansion

To asses the validity of the epsilon expansion, and so the stability of the fixed point we need to consider the behavior of all the marginal and relevant running couplings. Using Ward identities we can find directly the IR behavior for $\tilde{v}_{122}$:

$$\bar{v}_{122}(\rho) = \bar{v}_{11}(\rho)/\bar{\chi}_{1o}(\rho). \tag{11.95}$$

This means that

$$\begin{cases} \text{for } \epsilon = 0 & \bar{v}_{122} \sim 1/\ln\rho \\ \text{for } \epsilon > 0 & \bar{v}_{122} \sim \rho^{\epsilon/2}. \end{cases} \tag{11.96}$$

In the same way we can obtain the behavior of $\bar{v}_{2222}$

$$\bar{v}_{2222}(\rho) = 3\bar{v}_{122}(\rho)/\bar{\chi}_{1o}(\rho). \tag{11.97}$$

In this way:

$$\begin{cases} \text{for } \epsilon = 0 & \bar{v}_{2222} \sim 1/\ln\rho \\ \text{for } \epsilon > 0 & \bar{v}_{2222} \to \bar{v}^*_{2222} \end{cases} \tag{11.98}$$

From (11.98) and (11.96) we can draw one important conclusion: for $\epsilon = 0$ the theory is asymptotically free. Marginal interactions become marginally irrelevant (they vanish like $1/\ln\rho$) and the results obtained at one loop are exact.

Unluckily for $\epsilon > 0$ the theory is not free, we have a non trivial line of fixed points for the running coupling $\bar{v}_{2222}$ and so all results can be considered correct only at the lowest order in ϵ. We will show in Sec. 11.1.12 that these results are instead exact for $\epsilon < 2$. To demonstrate this fact we will study the whole renormalized perturbation theory. Note however that contrary to what happens generally, the asymptotic behavior of $\bar{v}_{11}$ does not depend on the position of the fixed point, but only on its very existence. This is clear from equation (11.97) and (11.95), if $\bar{v}_{2222}(\rho)$ is finite for $\rho \to 0$ then $\bar{v}_{11}(\rho) \sim \rho^{\epsilon}$. This means that the presence of a line of fixed points is not in contrast with universality, that actually takes place.

11110   Scaling of the Dimensional Vertex Functions

We have seen in (11.71) that, to obtain the asymptotic behavior of the vertex function, we must consider also its original dimension. This fact gives for our vertex parts the following solution. For $d = 3$ we have:

$$\begin{cases} \Gamma_{11}(k) & \sim & 1/\ln(\kappa/\Lambda) \\ \Gamma_{12}(k) & \sim & k_0/\ln(\kappa/\Lambda) \\ \Gamma_{22}(k) & \sim & k^2 \\ \Gamma_{122}(k_i)|_{SP(\kappa)} & \sim & 1/\ln(\kappa/\Lambda) \\ \Gamma_{2222}(k_i)|_{SP(\kappa)} & \sim & 1/\ln(\kappa/\Lambda)\,, \end{cases} \tag{11.99}$$

while for $d < 3$ the result is the following:

$$\begin{cases} \Gamma_{11}(k) & \sim & \kappa^\epsilon \\ \Gamma_{12}(k) & \sim & \kappa^\epsilon k_0 \\ \Gamma_{22}(k) & \sim & k^2 \\ \Gamma_{122}(k_i)|_{SP(\kappa)} & \sim & \kappa^\epsilon \\ \Gamma_{2222}(k_i)|_{SP(\kappa)} & \sim & \kappa^\epsilon\,. \end{cases} \tag{11.100}$$

so we find that the three vertex parts Γ_{11}, Γ_{122} and Γ_{2222} are scaling in the same way. This result comes of course from WI and is a particular case of a more general rule that will be discussed in details in the following Section.

It can be interesting at this point to make a comparison with the standard perturbation theory for bosons and write the normal and anomalous self energies (as defined for example in Ref. 35 $B(k) = \Sigma_{12}(k)$) in terms of the Γ_{ij}:

$$\begin{cases} -\mu + \Sigma_{11}(k) & = & [\Gamma_{22}(k) + \Gamma_{11}(k) + 2i\Gamma_{12}(k)]/4 \\ \Sigma_{12}(k) & = & [\Gamma_{11}(k) - \Gamma_{22}(k)]/4 \end{cases} \tag{11.101}$$

The results obtained above show that $\Sigma_{12}(k) \sim \Gamma_{11}(k)$ because $\Gamma_{22}(k)$ vanishes like k^2. So the result found for $\Gamma_{11}(k)$ is equivalent to the finding that $\Sigma_{12}(k = 0) = 0$ by NN [77]. In this way the Hughenoltz-Pines identity for $1 < d \le 3$ becomes:

$$\Sigma_{11}(0) = \mu \qquad \text{because} \qquad \Sigma_{12}(0) = 0, \tag{11.102}$$

but at the same time the sound mode remains unchanged.

11111   Generalization of the results for vertex func tions of higher order

With the help of WI written in their general form [cfr. (8.85)] we can find simple and strong conditions on the behavior of the higher order vertex functions. We consider (8.85) with $\chi_{2o} = 0$ and k appearing in the composite part equal to zero:

$$\Gamma_{n_1,n_2}\chi_{1o} + n_1\Gamma_{n_1-1,n_2} + (n_2 - 1)\Gamma_{n_1+1,n_2-2} = 0\,. \tag{11.103}$$

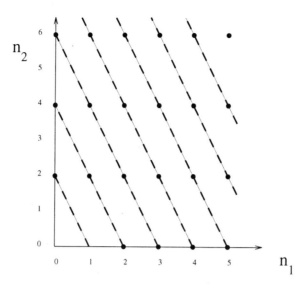

Figure 11.3: Lines of Γ with the same IR behavior

We suppose that Γ_{n_1-1,n_2} is less divergent than the other two Γ functions, in this way we can write asymptotically for $k \to 0$ that:

$$\Gamma_{n_1,n_2}\chi_{1o} + (n_2 - 1)\Gamma_{n_1+1,n_2-2} = 0. \qquad (11.104)$$

Equation (11.104) gives a relation among the Γ functions and relates all the non vanishing ones to $\Gamma_{n_1,0}$. In fact for odd number of 2 legs the vertex functions vanish by Time reversal, and successive use of (11.104) leads to:

$$\Gamma_{n_1,2n}(k_i) \sim \Gamma_{n_1+1,2n-2}(k_i) \sim \ldots \sim\sim \Gamma_{n_1+n,0}(k_i), \qquad (11.105)$$

when all the k_i are going to zero together.

As a matter of facts the scaling relations between v_{11}, v_{122}, and v_{2222} corresponds exactly to one of these chains of relations, and this explain the result found in (11.100). It is useful to represent pictorially these relations on the plane (n_1, n_2) of all the Γ_{n_1,n_2}. They connect vertex functions that are on the lines of equations $n_2 = 2N - 2n_1$ (see Fig. 11.3) where N is an integer. It is thus possible to identify each of this lines with a value of N.

We check now that the hypothesis that the singular behavior grows with N, as supposed to obtain (11.104). To this purpose we must find the IR behavior of all the $\Gamma_{n_1,0}$ (note that $N = n_1$ when $n_2 = 0$, so we can identify the whole line of Γ functions with the first $\Gamma_{N,0}$).

The simplest way to do this is to change slightly the RG approach and renormalize the longitudinal correlation functions with a wave function renormalization instead of using the v_{11} running coupling. This will keep the propagator fixed and the renormalization required will spread over all the Γ functions. We must discuss briefly the way in which this can be performed. We renormalize the functions with n_1 legs of type 1 and zero legs of type 2 by introducing the following factor:

$$\Gamma^R_{n_1,0}(k_i) = Z_1^{n_1/2}\Gamma_{n_1,0}(k_i) \qquad (11.106)$$

from this equation it is clear that $Z_1(\rho) \sim v_{11}(\rho)$ the RG equation can be modified as to introduce the anomalous exponent due to Z_1, in practice it only introduces a correction to the bare dimension of the vertex functions equal to $n_1\epsilon/2$. So that the IR behavior of these Γ functions becomes (cfr. (9.46)):

$$\Gamma^R_{n_1,0}(k) \sim k^{4-2n_1+n_1\epsilon} \qquad (11.107)$$

This gives for the first Γ functions the following behavior:

$$\begin{align}
\Gamma_{11}(k) &\sim k^\epsilon & (11.108)\\
\Gamma_{111}(k) &\sim k^{-2+2\epsilon} & (11.109)\\
\Gamma_{1111}(k) &\sim k^{-4+3\epsilon} & (11.110)\\
&\cdots & (11.111)
\end{align}$$

In this way we obtain that the hypothesis made on the IR behavior of the $\Gamma_{n_1,0}$ is correct:

$$\Gamma_{n_1,0}(k) < \Gamma_{n_1+1,0}(k) \qquad \text{for } k \to 0. \qquad (11.112)$$

We ask now how the RG process proceed for the Γ_{n_1,n_2}. By using the relation (11.105) we obtain that the correct renormalization for the vertex functions is given by the following expression:

$$\Gamma^R_{n_1,2n} = Z_1^{\frac{n_1+n}{2}} Z_1^{-n/2}\Gamma_{n_1,2n} = Z_1^{n_1/2}\Gamma_{n_1;2n} \qquad (11.113)$$

where the first term comes from the renormalization of the $\Gamma_{n_1+n,0}$ while the second comes from the change in the bare dimension between $\Gamma_{n_1+n,0}$ and $\Gamma_{n_1,2n}$. The final result shows that Z_1 is really a wave function renormalization for this system as it appears in each vertex function with the same multiplicity of the type 1 external legs.

We have thus found at one loop order the scaling behavior of all the non vanishing (even number of type 2 legs) vertex functions. We will not discuss the general scaling of the vertex functions with odd number of type 2 legs. In the following Section we discuss how the theory can be generalized to all orders in ϵ, and we will make use of the results for $\Gamma_{n1,2n}$ found here.

11112   Renormalized Perturbation Theory

As we have already said the scaling behavior found for the vertex functions is exact for $\epsilon = 0$, but it is apparently correct only at order ϵ for $\epsilon > 0$. Is this Section we give a simple argument on the renormalized perturbation

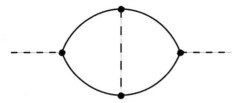

Figure 11.4: Two loops diagram for Γ_{11}

theory that leads to the generalization of the result at all orders in ϵ. In the next Section we will consider a different demonstration of the same result by studying the skeleton structure of the perturbation theory.

First of all we recognize that the renormalized Green's Functions behaves like the unrenormalized ones except for the $\mathcal{G}_{11}$ that changes drastically [cf (11.25)-(11.27)]:

$$
\begin{cases}
\mathcal{G}_{11}(k) &= \dfrac{\tilde{u}_{22}k_0^2 + \tilde{z}_{22}\mathbf{k}^2}{c_o^2\tilde{v}_{11}\tilde{z}_{22}}\dfrac{1}{k^2} \quad \sim \quad \kappa^{-\epsilon} \\[2ex]
\mathcal{G}_{12}(k) &= -\dfrac{c_o\tilde{w}_{12}}{\tilde{v}_{11}\tilde{z}_{22}}\dfrac{k_0}{k^2} \quad \sim \quad k_0/k^2 \\[2ex]
\mathcal{G}_{22}(k) &= \dfrac{1}{\tilde{z}_{22}}\dfrac{1}{k^2} \quad \sim \quad 1/k^2
\end{cases}
\tag{11.114}
$$

So the study of the perturbation theory that has been done is no more correct, due to the behaviour of the $\mathcal{G}_{11}$ that diverges like $\kappa^{-\epsilon}$. This is true also for the dimensionless coupling $\bar{v}_{11}$. So we need to reconsider the role of $\mathcal{G}_{11}$ in the perturbation theory. We begin by noting that the only non irrelevant coupling that can couple with $\mathcal{G}_{11}$ is $\bar{v}_{122}$. So in the internal part of the diagrams one always find the product of 3 terms:

$$
\bar{v}_{122}^2\mathcal{G}_{11} \sim \frac{\bar{v}_{122}^2}{\bar{v}_{11}} \sim 1
\tag{11.115}
$$

so the infinite contribution coming from $\mathcal{G}_{11}$ is always exactly compensated by the vanishing running coupling $\bar{v}_{122}$.

We consider the perturbation theory for $\tilde{v}_{11}$, in particular we consider the two loop diagram with an internal $\mathcal{G}_{11}$ line (see Fig.11.4).

This diagram will contribute with a term of the following form:

$$
\tilde{v}_{122}^2\bar{v}_{2222}^*\kappa^\epsilon\kappa^{-2\epsilon}.
\tag{11.116}
$$

The last term comes from the 2 loops of integration, the term $\tilde{v}_{122}$ corresponds to the two entering vertices for the diagram, and the term $\bar{v}_{2222}^*\kappa^\epsilon$ is the product of the Green's function with the two vertices.

We can see that the other terms in the expansion have a similar behavior and the RG equation for $\bar{v}_{11}$ shows the following form:

$$\bar{v}'_{11} = \bar{v}_{11} + \frac{\bar{v}_{11}^2}{\bar{\chi}_{1o}^2 \bar{z}_{22}^2} \left[A_0 + A_1 \bar{v}_{2222}^* + A_2 (\bar{v}_{2222}^*)^2 + \ldots \right] = \bar{v}_{11} + \frac{\bar{v}_{11}^2}{\bar{\chi}_{1o}^2 \bar{z}_{22}^2} A(\bar{v}_{2222}^*)$$

(11.117)

Remember that the scaling of $\bar{\chi}_{1o}(\rho)$ is known exactly and the Ward identities guaranty that $\bar{z}_{22}$ does not scale; so the RG equation becomes:

$$\rho \frac{d\bar{v}_{11}}{d\rho} = \frac{\bar{v}_{11}^2}{\bar{\chi}_{1o}(\rho)^2 \bar{z}_{22}^2} A(\bar{v}_{2222}^*)$$

(11.118)

and its solution is simply:

$$\bar{v}_{11}(\rho) = \frac{\bar{\chi}_{1o}(1)^2 \bar{z}_{22}^2 \epsilon}{A(\bar{v}_{2222}^*)} \rho^\epsilon$$

(11.119)

We will give a different demonstration of this fact in the next Section using the skeleton expression of the whole perturbation theory.

At this point it is possible to conclude that the *exact* IR behavior of the system has been found, *i.e.* that the behavior found at first order in ϵ is actually correct at all orders in ϵ. We will discuss in 11.3 in details the scaling and the form of the correlation functions, at this point we only want to summarize some points. The sound mode in the Green's functions is completely preserved by the WI and is completely independent from the IR behavior of the $\bar{v}_{11}$ running coupling. This last coupling shows a peculiar behavior depending strongly on dimensionality. In fact for $d > 3$ Bogolubov approximation gives the correct IR behavior of the system because no IR divergences are present, for $d = 3$ we find asymptotic freedom and the anomalous Self Energy goes to zero logarithmically, while for $1 < d < 3$ the IR behavior is controlled by a novel line of fixed point far apart from the Bogolubov gaussian fixed point. For $d = 1$ $(T = 0)$ our approach should be strongly modified due to the disappearance of the condensate.

11.2   Skeleton Structure of the Perturbation Theory

In this Section we study briefly the skeleton structure of the perturbation theory. The purpose is to clarify some of the results obtained in the previous Sections and to check that the irrelevant running couplings cannot spoil the solution found also if the epsilon expansion is extended down to $\epsilon = 2$. It turns out a simple picture of the mechanism with whom the diverging contributions sums up to give zero or finite results.

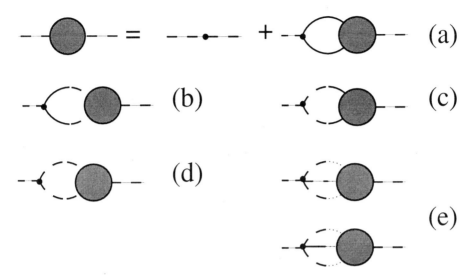

Figure 11.5: Diagrammatic equation for Γ_{11}, all the internal lines are full Green function. The dotted line stands for any of the two possible legs (*i.e.* dashed or full line).

We begin by considering only the vertex functions which by power counting are marginal at $d = 3$, we will then consider all the other Γ functions.

So we write the first equation for $\Gamma_{11}(k)$ (cfr. Fig. 11.5 (a)):

$$\Gamma_{11}(k) = v_{11} - v_{11} \sum_q \mathcal{G}_{22}(q)\mathcal{G}_{22}(q + k)\Gamma_{221}(q, -q - k) \tag{11.120}$$

the coupling appearing in this equation are the bare ones and no irrelevant running coupling and Γ functions are included. The Green functions appearing in this and the following are the exact ones. We have seen that the IR behavior of $\mathcal{G}_{22}$ and $\mathcal{G}_{12}$ is fixed exactly by WI.

Equation (11.120) can be easily solved by using the WI that links $\Gamma_{221}(q, -q - k)$ to Γ_{11}. The equation reduces to an algebraic equation for $k \to 0$ dominated by the divergent loop $(k^{-\epsilon})$. The result is simply:

$$\Gamma_{11}(k) = \frac{v_{11}}{1 + Ak^{-\epsilon}} \sim k^{\epsilon}. \tag{11.121}$$

where A is a constant. This result is equivalent to the statement that $\bar{v}_{11}(\rho) \sim \rho^{\epsilon}$ for any value of $\epsilon < 2$, found by discussing the renormalized perturbation theory.

We now consider the effect of the presence of the other vertex functions [see Fig. 11.5 (b)], in this case only Γ_{111}, we suppose for it that the behavior found by RG method is correct, *i.e.* $\Gamma_{111}(k) \sim k^{-2+2\epsilon}$ [hereafter we simplify the notation by indicating only one k as the argument of the $\Gamma(k_1, k_2 \ldots)$ because we are considering the limit $\lambda \to 0$ for $\Gamma(\lambda k_i)$], the full Green functions are constrained by WI so that $\mathcal{G}_{12}(k) \sim k_0/k^2$ this proves that the contribution of this diagram is finite (by power counting it has dimension ϵ), and will only change the coefficient in the scaling behavior.

Lastly we investigate the contribution of an irrelevant running coupling like v_{111} [see Fig. 11.5 (c)]. This irrelevant running coupling can couple with a loop with 2 $\mathcal{G}_{12}$ to Γ_{122}. By power counting this loop (only the loop) is finite as it goes like $\epsilon - 2$, but one may wonder if it can change the IR behavior due to the fact that it is multiplied by Γ_{122} ($= \Gamma_{11}$). The equation can be rewritten explicitly in the following way:

$$\Gamma_{11}(k) = B + (Ak^{-\epsilon} + C)\Gamma_{11}(k) \tag{11.122}$$

where C is the contribution of this loop and B the contribution coming from the loop with Γ_{111}. It is then clear that the irrelevant running couplings do are effectively irrelevant because they change only the value of the coefficient:

$$\Gamma_{11}(k) = \frac{B}{1 - C - Ak^{-\epsilon}} \sim \frac{B}{A}k^{\epsilon}, \tag{11.123}$$

The other irrelevant running coupling that appears in the bare Action are v_{1111}, v_{1122} [see Fig. 11.5 (e)]. It is easy to see by explicit power counting that they always lead to finite corrections in a similar way to v_{111}. The only new point here is that one need to use the scaling of Γ_{11} and so of $\mathcal{G}_{11}$ found above.

To find these result we had to use the IR behavior of $\Gamma_{111}(k)$. It is possible to write a skeleton equation also for Γ_{111} and to solve it. Again the irrelevance of the other terms in this equation is demonstrated by assuming the IR behavior for Γ_{1111}. The procedure can be performed in general for all the Γ functions, to elucidate this procedure we begin by considering the equation for Γ_{111} because new features appear with respect to the equation for Γ_{11}. In fact the leading contributions to this equation comes from two terms (see Fig. 11.6) one with $\Gamma_{2211} \sim \Gamma_{111}$ and an other with two Γ_{122}. An explicit expression of this equation with all the divergences explicitly written is the following:

$$\Gamma_{111}(k) = k^{-\epsilon}\Gamma_{111}(k) + k^{-2+\epsilon} \tag{11.124}$$

so that the solution reads:

$$\Gamma_{111}(k) = \frac{k^{-2+\epsilon}}{1 - k^{-\epsilon}} \sim k^{-2+2\epsilon}. \tag{11.125}$$

It is then clear how it is possible that the finite terms never spoil the IR behavior found with the RG, in fact the IR behavior comes out as a ratio of two diverging quantities. The structure of the solution also enlights the mechanism that lead to the vanishing or the divergence of this vertex function.

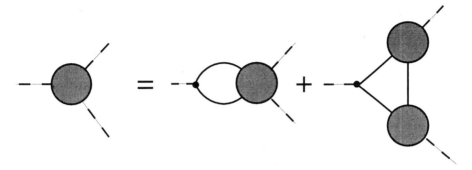

Figure 11.6: Equation for Γ_{111}. Again all the lines are exact Green functions.

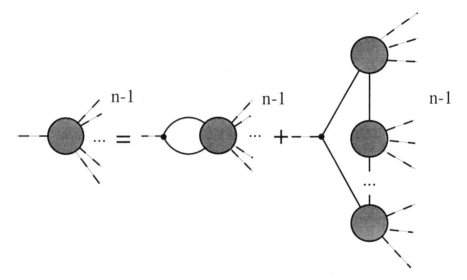

Figure 11.7: Equation for $\Gamma_{n,0}$. All the internal lines are full $\mathcal{G}_{22}$.

The equation found for Γ_{111} can be easily generalized for vertex functions with arbitrary number of legs of type 1. The generalized equation is shown in Fig. 11.7. Here we are considering the contribution to $\Gamma_{n,0}$ (we are using here the notation introduced in. 8.2.3) coming from all the "most relevant" Γ functions $i.e.$ by $\Gamma_{n-1,2}, \Gamma_{n-2,2}, \ldots, \Gamma_{1,2}$. The argument that these are the most relevant comes from the fact that we need always two internal legs of type 2 to have the most diverging correlation functions $\mathcal{G}_{22}$ and any number of external 1 legs. $\Gamma_{n-1,2}$ using (11.104) is related to $\Gamma_{n,0}$, while the other diagrams in Fig. 11.7 lead to diverging contributions. The important feature is that the power of divergence is exactly the same for each of these diagrams and can be readily evaluated to be $k^{(\epsilon-2)(n-2)}$ so that the equation for $\Gamma_{n,0}$ acquires a form similar

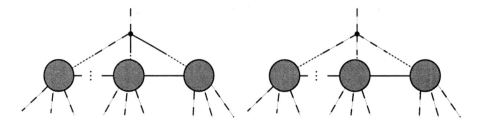

Figure 11.8: Diagrams that contains irrelevant running coupling. The dotted lines indicate all the possible correlation functions.

to (11.124):

$$\Gamma_{n,0}(k) = Ak^{-\epsilon}\Gamma_{n,0} + Bk^{(\epsilon-2)(n-2)} \qquad (11.126)$$

again we find the sum of two terms, both diverging for $d > 1$ thus leading to the following expression for $\Gamma_{n,0}$:

$$\Gamma_{n,0}(k) = \frac{Bk^{(\epsilon-2)(n-2)}}{1 - Ak^{-\epsilon}} \sim k^{(\epsilon-2)(n-2)+\epsilon} \qquad \text{for } n > 1. \qquad (11.127)$$

that gives the same behavior found with the RG method for all these Γ functions [cfr. (11.107)]. Once the behavior of all the $\Gamma_{n_1,0}$ is known by making use of (11.104) it is possible to know the behavior of all the other $\Gamma_{n_1,2n}$. The vertex functions with odd number of 2 legs vanish in the lowest order, *i.e.* they begin with some power of ω and are not discussed.

We are left with the following question: do the other Γ functions not considered in Fig. (11.7) change this simple picture? The answer is no, and the demonstration follows the lines of the demonstration that Γ_{111} contribute with only finite terms to Γ_{11}. In fact, it is possible to show that the change of the type (from 2 to 1) of any of the legs in that diagrams lower the power of divergence by a factor k^{ϵ} this introduces subleading terms (with respect the B terms) in (11.126) that do not change the result.

The introduction of irrelevant terms gives an analogous behavior also if the discussion is a bit complicated by the appearance of two-loop diagrams (see Fig. (11.8)). In any case it is possible to show that diagrams that contains irrelevant running coupling are subleading. Note that these checks that within our treatment appears to be quite simple are in general very difficult to perform, see for instance the case of the introduction of a ϕ^6 term in the action for the ϕ^4 model that is considered in details in Sec. II.2 of Ref. 98. Note also that the fact that some of the running coupling that are irrelevant for $d = 3$ becomes relevant for lower d does not change the discussion, as all divergences found remains the leading ones for $\epsilon < 2$. So the the running couplings dimensionless at $d = 3$ control the IR behavior also for $d < 3$.

In conclusion we have seen that the RG group gives a guide on the effective relevance of the terms appearing in the skeleton structure of the theory. In this way it is possible to solve the equations for the most relevant terms and discuss at a second time that the other terms do not spoil the solution.

The form of the solution given for the vertex functions shows clearly that it is not possible to recover the same solution by finite order perturbative calculation. At the same time it gives a simple way in which one can see how the singular series sums up. Consider for instance $\Gamma_{11}(k)$:

$$\Gamma_{11}(k) = \frac{A}{1 - Bk^{-\epsilon}} = \sum_n B^n k^{-n\epsilon} . \tag{11.128}$$

It is then possible a comparison with the general form given by (10.42). In this case no regular term appears as in fact the $k \to 0$ value is 0.

11.3 Solutions for the $\langle \psi\psi \rangle$ and $\langle J_\mu J_\nu \rangle$ Correlations Functions

We now "convert" the RG flow equations in the explicit form for the vertex functions and the Green's functions. In this way we describe completely the IR behavior of the Bose system. The fixed point is defined by the value of few physical quantities, specifically n_s, n_0, $dn/d\mu$, and $dn_0/d\mu$. The result of Gavoret and Nozieres is recovered at the leading order. Next-to-leading order contributions are singular in $1 < d \le 3$ and they lead to the vanishing of the anomalous self energy [114]. By contrast, the density-density, density-current, and current-current response functions turn out to be free from divergences.

It is worth saying that several attempts have been made to obtain consistent approximate results for the correlation functions and the response functions [116–125]. The old-standing problem of having "conserving" or "gap-less" approximations has in fact prevented the development of a self-consistent theory as in superconducting Fermi systems [28]. The results obtained here are consistent with the sum rules and WI for this system as they are obtained exploiting fully the local gauge symmetry of the theory.

1131   Traditional Greens Functions in Terms of the $\mathcal{G}_{ij}$

We begin with the definition of the usual correlation functions:

$$\begin{cases} \mathcal{G}(x,y) \equiv \langle \psi(x)\psi^*(y) \rangle_C \\ \mathcal{G}_1(x,y) \equiv \langle \psi(x)\psi(y) \rangle_C . \end{cases} \tag{11.129}$$

So in terms of our Green's functions we obtain [cf. (8.6) and (8.7)]:

$$\mathcal{G}(x,y) = \mathcal{G}_{11}(x,y) + \mathcal{G}_{22}(x,y) - i\mathcal{G}_{12}(x,y) + i\mathcal{G}_{21}(x,y), \tag{11.130}$$

and we have:

$$\mathcal{G}(p) = \mathcal{G}_{11}(p) + \mathcal{G}_{22}(p) - i\mathcal{G}_{12}(p) + i\mathcal{G}_{21}(p)..\tag{11.131}$$

Note that this relation holds also when the system is in a state of broken symmetry due to the fact that the Green's function are the connected ones.

In a similar way we can write the expression in terms of our Green's functions for the anomalous Green's function:

$$\mathcal{G}_1(x,y) = \mathcal{G}_{11}(x,y) - \mathcal{G}_{22}(x,y) + i\mathcal{G}_{12}(x,y) + i\mathcal{G}_{21}(x,y),\tag{11.132}$$

and the Fourier Transform is:

$$\mathcal{G}_1(p) = \mathcal{G}_{11}(p) - \mathcal{G}_{22}(p) + i\mathcal{G}_{12}(p) + i\mathcal{G}_{21}(p)..\tag{11.133}$$

1132 The Γ_{ij} in Terms of the Running Couplings

The v_{11} Running Coupling

We consider the fixed-point solution for the running coupling $\bar{v}_{11}$ [see (11.80)]:

$$\tilde{v}_{11}(\rho) = K_{4-\epsilon}^{-1}\bar{v}_{11}(\rho) \to -2K_{4-\epsilon}^{-1}\bar{\chi}_{1o}(1)^2\bar{z}_{22}(1)^2 \times \begin{cases} \frac{1}{\ln(\rho)} & (\epsilon = 0) \\ -\epsilon\rho^\epsilon & (0 < \epsilon < 2) \end{cases},\tag{11.134}$$

but $\bar{z}_{22}$ does not scale and we know that $z_{22} \to 2n_s/\chi_{1o}^2$, so we have the following simple and exact identity:

$$z_{22} = \tilde{z}_{22} = \bar{z}_{22}(\rho) = \bar{z}_{22}(1) = \frac{2n_s}{\chi_{1o}^2}.\tag{11.135}$$

We can find a similarly simple and exact relation for χ_{1o} [see (11.43) and (11.75)]:

$$\chi_{1o} = c_o^{3/2}\tilde{\chi}_{1o} = c_o^{3/2}\kappa^{-\epsilon/2}\bar{\chi}_{1o}(\rho),\tag{11.136}$$

so that:

$$\bar{\chi}_{1o}(1) = \frac{\chi_{1o}\kappa^{\epsilon/2}}{c_o^{3/2}}.\tag{11.137}$$

We now make use of the result (11.71) to obtain the scaling expression of $\bar{\Gamma}_{11}$:

$$\begin{aligned}\bar{\Gamma}_{11}^R(\rho\kappa,\underline{u},\kappa) &= c_o^2\bar{v}_{11}(\rho) \\ &= 2K_{4-\epsilon}^{-1}c_o^2\bar{\chi}_{1o}^2(1)\bar{z}_{22}(1)^2\epsilon\rho^\epsilon \\ &= 2K_{4-\epsilon}^{-1}c_o^2\frac{\chi_{1o}^2\kappa^\epsilon}{c_o^3}\left(\frac{2n_s}{n_0}\right)^2\epsilon\rho^\epsilon \\ &= \frac{8n_s^2\kappa^\epsilon}{K_{4-\epsilon}c_o n_0}\epsilon\rho^\epsilon.\end{aligned}\tag{11.138}$$

So that an explicit expression for $\bar{\Gamma}_{11}$ is the following:

$$\bar{\Gamma}_{11}(k) = \frac{8n_s^2 k^\epsilon \epsilon}{K_{4-\epsilon} c n_0} \qquad (0 < \epsilon < 2), \qquad (11.139)$$

Note that for $d = 3$ the expression for $\bar{\Gamma}_{11}$ simplifies somewhat:

$$\bar{\Gamma}_{11}(k) = -\frac{64\pi^2 n_s^2}{c_o n_0 \ln(k/\kappa)} \qquad (\epsilon = 0). \qquad (11.140)$$

The w_{12} Running Coupling

From the RG equations we know that:

$$\frac{\bar{w}_{12}(\rho)}{\bar{v}_{11}(\rho)} = \frac{\bar{w}_{12}(1)}{\bar{v}_{11}(1)}, \qquad (11.141)$$

moreover from (10.80) we get:

$$\frac{\bar{w}_{12}}{\bar{v}_{11}} \to -\frac{c_o^2}{\chi_{1o}} \frac{\partial\chi_{1o}}{\partial\mu}\bigg|_\lambda = -\frac{c_o^2}{2\,n_0} \frac{dn_0}{d\mu}\bigg|_\lambda \equiv -\frac{c_o^2}{2\,n_0} \frac{dn_0}{d\mu}. \qquad (11.142)$$

So for the bare dimensionless running couplings we have:

$$\frac{\bar{w}_{12}(1)}{\bar{v}_{11}(1)} = \frac{c_o^2}{2\,n_0} \frac{dn_0}{d\mu}. \qquad (11.143)$$

This means that:

$$\bar{w}_{12}(\rho) = \frac{\bar{w}_{12}(1)}{\bar{v}_{11}(1)} \bar{v}_{11}(\rho) = -\frac{c_o^2}{2\,n_0} \frac{dn_0}{d\mu} \bar{v}_{11}(\rho), \qquad (11.144)$$

and finally for $\bar{\Gamma}_{12}$:

$$\bar{\Gamma}_{12}(\mathbf{k},\omega) = \begin{cases} \dfrac{32\pi^2 n_s^2}{n_0^2 c} \dfrac{dn_0}{d\mu} \dfrac{1}{\ln(k/\kappa)}\omega & (\epsilon = 0) \\[2ex] -\dfrac{4n_s^2}{K_{4-\epsilon} c_o\, n_0^2} \dfrac{dn_0}{d\mu} \epsilon k^\epsilon \omega & (0 < \epsilon < 2) \end{cases} \qquad (11.145)$$

The u_{22} and z_{22} Running Couplings

Now we consider $\bar{\Gamma}_{22}$:

$$u_{22} = \frac{1}{c_o^2}\tilde{u}_{22} = \frac{1}{c_o^2}\bar{u}_{22} \to -\frac{1}{n_0} \frac{\partial^2\Gamma}{\partial\mu^2}\bigg|_{n_0} \frac{1}{\beta\Omega}, \qquad (11.146)$$

where use has been made of (10.79). In this way by using again the fixed-point value for $\bar{u}_{22}$ we get (cf. (11.76) in the limit $v_{11}(\rho) \to 0$):

$$\frac{1}{c^2}\left[\bar{u}_{22}(1) + \frac{\bar{w}_{12}(1)^2}{\bar{v}_{11}(1)}\right] = -\frac{1}{n_0}\frac{\partial^2\Gamma}{\partial\mu^2}\bigg|_{n_0}\frac{1}{\beta\Omega}. \qquad (11.147)$$

Note that in this case we have demonstrated by the one-loop calculation that the exact compressibility is finite (at least, for a low density or weak coupling Bose gas [cf. (10.77)]).

The expression of $\bar{u}_{22}$ in terms of physical derivatives is the following [cf. (11.78)]:

$$\bar{u}_{22}(\rho) = -\frac{c_o^2}{n_0}\frac{\partial^2\Gamma}{\partial\mu^2}\bigg|_{n_0}\frac{1}{\beta\Omega} - \left(\frac{c^2}{2n_0}\frac{dn_0}{d\mu}\right)^2\bar{v}_{11}(\rho).$$
(11.148)

So that we have:

$$\bar{u}_{22}(\rho) = -\frac{c_o^2}{n_0}\frac{\partial^2\Gamma}{\partial\mu^2}\bigg|_{n_0}\frac{1}{\beta\Omega} + \frac{c_o n_s^2}{n_0^3}\left(\frac{dn_0}{d\mu}\right)^2 \times \begin{cases} \dfrac{16\pi^2}{\ln(\rho)} & (\epsilon = 0) \\[2mm] -\dfrac{2\kappa^\epsilon}{K_{4-\epsilon}}\epsilon\rho^\epsilon & (0 < \epsilon < 2) \end{cases}$$
(11.149)

Note also that:

$$c^2 = \frac{v_{11}z_{22}}{v_{11}u_{22} + w_{12}^2} \rightarrow \frac{z_{22}}{u_{22}} = -2n_s/\frac{\partial^2\Gamma}{\partial\mu^2}\bigg|_{n_0}\frac{1}{\beta\Omega},$$
(11.150)

so that

$$\frac{1}{\beta\Omega}\frac{\partial^2\Gamma}{\partial\mu^2}\bigg|_{n_0} = -2\frac{n_s}{c^2},$$
(11.151)

and the final expression for Γ_{22} becomes:

$$\bar{\Gamma}_{22}(\mathbf{k},\omega) = \begin{cases} \left[\dfrac{2n_s c_o^2}{n_0\, c^2} + \dfrac{16\pi^2 n_s^2 c_o}{n_0^3}\left(\dfrac{dn_0}{d\mu}\right)^2\dfrac{1}{\ln(k/\kappa)}\right]\dfrac{\omega^2}{c_o^2} + \dfrac{2n_s}{n_0}\mathbf{k}^2 & (\epsilon = 0) \\[4mm] \left[\dfrac{2n_s c_o^2}{n_0\, c^2} - \dfrac{2n_s^2 c_o}{K_{4-\epsilon}n_0^3}\left(\dfrac{dn_0}{d\mu}\right)^2\epsilon k^\epsilon\right]\dfrac{\omega^2}{c_o^2} + \dfrac{2n_s}{n_0}\mathbf{k}^2 & (0 < \epsilon < 2) \end{cases}$$
(11.152)

In this way we have an explicit expression for the matrix $\bar{\Gamma}_{ij}$, which is valid in the IR asymptotic limit. In the three dimensional case the expression is the following:

$$\bar{\Gamma}_{ij}(\mathbf{k},\omega) = \begin{pmatrix} -\dfrac{64\pi^2 n_s^2}{c_o\, n_0\ln(k/\kappa)}, & \dfrac{32\pi^2 n_s^2}{c_o\, n_0^2}\dfrac{dn_0}{d\mu}\dfrac{\omega}{\ln(k/\kappa)} \\[4mm] -\dfrac{32\pi^2 n_s^2}{c_o\, n_0^2}\dfrac{dn_0}{d\mu}\dfrac{\omega}{\ln(k/\kappa)}, & \dfrac{2n_s}{n_0 c^2}(\omega^2 + c^2\mathbf{k}^2) + \dfrac{16\pi^2 n_s^2}{n_0^3 c_o}\left(\dfrac{dn_0}{d\mu}\right)^2\dfrac{\omega^2}{\ln(k/\kappa)} \end{pmatrix}$$
(11.153)

For $1 < d < 3$ we have instead:

$$\bar{\Gamma}_{ij}(\mathbf{k},\omega) = \begin{pmatrix} \dfrac{8n_s^2 \epsilon k^\epsilon}{K_{4-\epsilon} c_o \, n_0} & -\dfrac{4n_s^2 \epsilon k^\epsilon}{K_{4-\epsilon} c_o \, n_0^2} \dfrac{dn_0}{d\mu}\omega \\ \dfrac{4n_s^2 \epsilon k^\epsilon}{K_{4-\epsilon} c_o \, n_0^2} \dfrac{dn_0}{d\mu}\omega, & \dfrac{2n_s}{n_0 \, c^2}(\omega^2 + c^2 \mathbf{k}^2) - \dfrac{2n_s^2 \epsilon k^\epsilon}{K_{4-\epsilon} n_0^3 c_o}\left(\dfrac{dn_0}{d\mu}\right)^2 \omega^2 \end{pmatrix}.$$

$$(11.154)$$

By inverting (11.153) (or (11.154)) we can find the correlation functions $\mathcal{G}_{ij}$ for $k \to 0$.

1133   Result for the $\mathcal{G}_{ij}$ Greens Functions

For the connected single-particle Green's function of the theory we obtain the following expression depending on the spatial dimension. For $d = 3$ we obtain:

$$\mathcal{G}_{ij}(\mathbf{k},\omega) = \frac{1}{D(\mathbf{k},\omega)} \times$$
$$\begin{pmatrix} \dfrac{2n_s}{n_0 c^2}(\omega^2 + c^2 \mathbf{k}^2) + \dfrac{16\pi^2 n_s^2}{n_0^3 c_o}\left(\dfrac{dn_0}{d\mu}\right)^2 \dfrac{\omega^2}{\ln(k/\kappa)}, & -\dfrac{32\pi^2 n_s^2}{c_o \, n_0^2}\dfrac{dn_0}{d\mu}\dfrac{\omega}{\ln(k/\kappa)} \\ \dfrac{32\pi^2 n_s^2}{c_o \, n_0^2}\dfrac{dn_0}{d\mu}\dfrac{\omega}{\ln(k/\kappa)}, & -\dfrac{64\pi^2 n_s^2}{c_o \, n_0 \ln(k/\kappa)} \end{pmatrix},$$

$$(11.155)$$

while for $1 < d < 3$

$$\mathcal{G}_{ij}(\mathbf{k},\omega) = \frac{1}{D(k)} \times$$
$$\begin{pmatrix} \dfrac{2n_s}{n_0 \, c^2}(\omega^2 + c^2 \mathbf{k}^2) - \dfrac{2n_s^2 \epsilon k^\epsilon}{K_{4-\epsilon} n_0^3 c_o}\left(\dfrac{dn_0}{d\mu}\right)^2 \omega^2, & \dfrac{4n_s^2 \epsilon k^\epsilon}{K_{4-\epsilon} c_o \, n_0^2}\dfrac{dn_0}{d\mu}\omega \\ -\dfrac{4n_s^2 \epsilon k^\epsilon}{K_{4-\epsilon} c_o \, n_0^2}\dfrac{dn_0}{d\mu}\omega, & \dfrac{8n_s^2 \epsilon k^\epsilon}{K_{4-\epsilon} c_o \, n_0} \end{pmatrix}$$

$$(11.156)$$

Here:

$$D(\mathbf{k},\omega) = \begin{cases} -\dfrac{128\pi^2 n_s^3}{c^2 c_o n_0^2 \ln(k/\kappa)}(\omega^2 + c^2 \mathbf{k}^2) & (\epsilon = 0) \\ \dfrac{16n_s^3}{c^2 c_o n_0^2}(\omega^2 + c^2 \mathbf{k}^2)\dfrac{\epsilon k^\epsilon}{K_{4-\epsilon}} & (0 < \epsilon < 2) \end{cases}$$

$$(11.157)$$

Note that the terms in $1/\ln(k)^2$ (or $k^{2\epsilon}$) simplify exactly. So we can summarize:

$$\mathcal{G}_{22}(\mathbf{k},\omega) \;=\; \frac{c^2 n_0}{2n_s}\frac{1}{\omega^2 + c^2 \mathbf{k}^2}$$

$$(11.158)$$

$$\mathcal{G}_{12}(\mathbf{k},\omega) = \frac{c^2}{4n_s}\frac{dn_0}{d\mu}\frac{\omega}{\omega^2 + c^2\mathbf{k}^2} \tag{11.159}$$

$$\mathcal{G}_{11}(\mathbf{k},\omega) = -\frac{c^2}{8n_0 n_s}\left(\frac{dn_0}{d\mu}\right)^2 \frac{\omega^2}{\omega^2 + c^2\mathbf{k}^2} - \begin{cases} \dfrac{c_o\,n_0}{64\pi^2 n_s^2}\ln(k/\Lambda) & (\epsilon = 0) \\[2ex] -\dfrac{c_o n_0 K_{4-\epsilon}}{8n_s^2 \epsilon k^\epsilon} & (0 < \epsilon < 2) \end{cases} \tag{11.160}$$

Remarkably only the last term depends on dimensionality.

1134   Ordinary Greens Functions

The Green's Function in the ψ-representation can be obtained in a simple way from (11.131) and (11.133) by substituting (11.158)-(11.160):

$$\begin{cases} \mathcal{G}(p) = \dfrac{m\,c^2 n_0}{n_s}\dfrac{1}{\omega^2 + c^2\mathbf{k}^2} - \dfrac{m\,c^2}{n_s}\dfrac{dn_0}{d\mu}\dfrac{i\,\omega}{\omega^2 + c^2\mathbf{k}^2} + \mathcal{G}_{11}(k) \\[2ex] \mathcal{G}_1(p) = -\dfrac{m\,c^2 n_0}{n_s}\dfrac{1}{\omega^2 + c^2\mathbf{k}^2} + \mathcal{G}_{11}(k) \end{cases} \tag{11.161}$$

(we have restored the canonical dimension by reintroducing the mass m) so that in the IR we have recovered the Gavoret and Nozieres result plus the sub-leading logarithmic (or $k^{-\epsilon}$) divergence and finite or less singular terms [126].

1135   Composite Greens Functions

We turn our attention now to the response functions. To obtain the composite Green's function $\mathcal{G}_{;\mu\nu}(p)$ we need first their expression in terms of the composite vertex parts [cf. Section 9.2.3, in particular (9.113)]:

$$\Gamma_{;\mu} = -\mathcal{G}_{;\mu}. \tag{11.162}$$

By differentiating with respect to A_ν at fixed χ_{oi} we obtain:

$$\begin{aligned} \Gamma_{;\mu\nu} &= -\mathcal{G}_{;\mu\nu} - \frac{\delta\mathcal{G}_{;\mu}}{\delta\lambda_i}\frac{\delta\lambda_i}{\delta A_\nu} \\ &= -\mathcal{G}_{;\mu\nu} - \mathcal{G}_{i;\mu}\Gamma_{i;\nu}. \end{aligned} \tag{11.163}$$

By writing it explicitly we get:

$$\Gamma_{;\mu\nu}(x,y) = -\mathcal{G}_{;\mu\nu}(x,y) - \int dx'\mathcal{G}_{i;\mu}(x',x)\Gamma_{i;\nu}(x',y), \tag{11.164}$$

and by use of (9.128):

$$\mathcal{G}_{;\mu\nu}(x,y) = -\Gamma_{;\mu\nu}(x,y) + \int dx'\int dy'\mathcal{G}_{ii'}(x',y')\Gamma_{i';\mu}(y',x)\Gamma_{i;\nu}(x',y). \tag{11.165}$$

The Fourier transform of $\mathcal{G}_{;\mu\nu}$ is given by the following expression (where use has been done of (9.68)):

$$\mathcal{G}_{;\mu\nu}(p) = -\Gamma_{;\mu\nu}(p) - \Gamma_{i';\mu}(-p)\mathcal{G}_{i'i}(p)\Gamma_{i;\nu}(p) \,. \tag{11.166}$$

[See (9.107) for the expressions that take into account the symmetry of the vertex functions $\Gamma_{i;\mu}$ and (9.109) for the power counting of the $\mathcal{P}$ functions.]

Density-Density Response Function

With the help of the Ward identities [see (10.51) and (10.54)] we can write expressions for the composite vertex functions that depend on the running couplings:

$$\Gamma_{2;0}(\mathbf{k}, \omega) = u_{22}n_0^{1/2}\omega \tag{11.167}$$

$$\Gamma_{1;0}(\mathbf{k}, \omega) = w_{12}n_0^{1/2} \,. \tag{11.168}$$

The Green functions can be written in the following form:

$$\mathcal{G}_{11}(p) = \frac{\Gamma_{22}(p)}{D(p)} \tag{11.169}$$

$$\mathcal{G}_{12}(p) = -\frac{\Gamma_{12}(p)}{D(p)} \tag{11.170}$$

$$\mathcal{G}_{22}(p) = \frac{\Gamma_{11}(p)}{D(p)} \tag{11.171}$$

$$D(p) = \Gamma_{11}(p)\Gamma_{22}(p) - \Gamma_{12}(p)\Gamma_{21}(p) \,. \tag{11.172}$$

The vertex parts are simply given by:

$$\Gamma_{22} = u_{22}\omega^2 + z_{22}\mathbf{k}^2, \qquad \Gamma_{11} = v_{11}, \qquad \Gamma_{12} = -\Gamma_{21} = w_{12}\omega \,, \tag{11.173}$$

and the explicit expression for $\mathcal{G}_{00}$ is the following [cf. (11.166)]:

$$
\begin{aligned}
\mathcal{G}_{;00}(p) &= -\Gamma_{;00}(p) - \Gamma_{1;0}(-p)\mathcal{G}_{11}(p)\Gamma_{1;0}(p) - \Gamma_{1;0}(-p)\mathcal{G}_{12}(p)\Gamma_{2;0}(p) \\
&\quad -\Gamma_{2;0}(-p)\mathcal{G}_{21}(p)\Gamma_{1;0}(p) - \Gamma_{2;0}(-p)\mathcal{G}_{22}(p)\Gamma_{2;0}(p) \\
&= -\Gamma_{;00}(p) - \Gamma_{1;0}(p)^2\mathcal{G}_{11}(p) - 2\Gamma_{1;0}(p)\Gamma_{2;0}(p)\mathcal{G}_{12}(p) + \Gamma_{2;0}(p)^2\mathcal{G}_{22}(p) \\
&= -\Gamma_{;00}(p) + \frac{-\Gamma_{1;0}(p)^2\Gamma_{22}(p) + 2\Gamma_{1;0}(p)\Gamma_{2;0}(p)\Gamma_{12}(p) + \Gamma_{2;0}(p)^2\Gamma_{11}(p)}{D(p)} \\
&= -\Gamma_{;00}(p) + \frac{n_0}{D(p)}\left[-w_{12}^2(u_{22}\omega^2 + z_{22}\mathbf{k}^2) + 2w_{12}^2u_{22}\omega^2 + v_{11}u_{22}^2\omega^2\right] \\
&= -\Gamma_{;00}(p) + \frac{n_0}{D(p)}\left[u_{22}\omega^2(v_{11}u_{22} + w_{12}^2) - w_{12}^2z_{22}\mathbf{k}^2\right] \,, \tag{11.174}
\end{aligned}
$$

where for $D(p)$ we have the following expression:

$$D(p) = (v_{11}u_{22} + w_{12}^2)\omega^2 + v_{11}z_{22}\mathbf{k}^2 \,. \tag{11.175}$$

Now we write $\mathcal{G}_{;00}$ in the following way:

$$
\begin{aligned}
\mathcal{G}_{;00}(p) &= -\Gamma_{;00} + \frac{n_0 u_{22}}{\mathcal{D}}\left\{\omega^2(v_{11}u_{22} + w_{12}^2) - \frac{w_{12}^2 z_{22}}{u_{22}}\mathbf{k}^2 + v_{11}z_{22}(\mathbf{k}^2 - \mathbf{k}^2)\right\} \\
&= -\Gamma_{;00} + \frac{n_0 u_{22}}{\mathcal{D}}\left\{\mathcal{D} - \left(\frac{w_{12}^2 z_{22}}{u_{22}} + v_{11}z_{22}\right)\mathbf{k}^2\right\} \\
&= -\Gamma_{;00} + n_0 u_{22} - n_0 z_{22}\frac{\mathbf{k}^2}{\omega^2 + c^2\mathbf{k}^2} \\
&= -\frac{2n_s \mathbf{k}^2}{\omega^2 + c^2\mathbf{k}^2},
\end{aligned}
\tag{11.176}
$$

Where we made use of (10.67) and (10.69). These expressions coincide with the ones given by Gavoret and Nozières [82], apart from a sign due to a different definition.

Current Response Functions

The expression of the Vertex functions in terms of the running couplings is obtained again from the Ward identities and symmetry considerations:

$$
\begin{aligned}
\Gamma_{2;l} &= i\mathbf{k}_l z_{22} n_0^{1/2} \tag{11.177}\\
\Gamma_{1;l} &= 0. \tag{11.178}
\end{aligned}
$$

The last expression should be written more explicitly:

$$
\Gamma_{1;l} = i\mathbf{k}_l \omega \mathcal{P}_{1;v}. \tag{11.179}
$$

But it appears always in combination with at least terms going to zero as k, so it would contribute always sub-leading terms. In this way we can write the explicit expression for the current-current and current-density response functions [cf. (11.166)]:

$$
\begin{aligned}
\mathcal{G}_{;0l}(p) &= -\Gamma_{;0l}(p) - \Gamma_{i';0}(-p)\mathcal{G}_{i'i}(p)\Gamma_{i;l}(p) \tag{11.180}\\
\mathcal{G}_{;ml}(p) &= -\Gamma_{;ml}(p) - \Gamma_{i';m}(-p)\mathcal{G}_{i'i}(p)\Gamma_{i;l}(p). \tag{11.181}
\end{aligned}
$$

We begin with the current-current functions [cf. (11.178)]:

$$
\begin{aligned}
\mathcal{G}_{;ml}(p) &= -\Gamma_{;ml}(p) - \Gamma_{2;m}(-p)\mathcal{G}_{22}(p)\Gamma_{2;l}(p) \\
&= -\Gamma_{;ml}(p) - (-i\mathbf{k}_m)z_{22}n_0^{1/2}\frac{v_{11}}{(v_{11}u_{22} + w_{12}^2)\omega^2 + v_{11}z_{22}\mathbf{k}^2}z_{22}n_0^{1/2}(i\mathbf{k}_l) \\
&= -\Gamma_{;ml}(p) - \frac{2n_s c^2 \mathbf{k}_m \mathbf{k}_l}{\omega^2 + c^2\mathbf{k}^2}. \tag{11.182}
\end{aligned}
$$

We use the representation of $\Gamma_{;ml}$ given by (10.62). We have seen that the longitudinal part of $\Gamma_{;ml}$ is proportional to the superfluid density [see 10.69].

We can substitute in the expression for $\mathcal{G}_{;ml}$ and obtain:

$$\mathcal{G}_{;ml}(k) = -\left(\delta_{ml} - \frac{k_m k_l}{\mathbf{k}^2}\right)\phi_t - 2n_s k_m k_l \left[\frac{c^2}{\omega^2 + c^2 \mathbf{k}^2} + \frac{1}{\mathbf{k}^2}\right]. \qquad (11.183)$$

Recall that the usual definition of the current-current response function χ_{lm} is the following:

$$\chi_{ml}(p) = -\mathcal{G}_{;ml}(p) - 2n\delta_{ml}, \qquad (11.184)$$

where the last term is the diamagnetic current response. In this way we obtain for χ_{ml}:

$$\chi_{ml}(k) = -\frac{k_m k_l}{\mathbf{k}^2}\phi_t + 2n_s k_m k_l \left[\frac{c^2}{\omega^2 + c^2 \mathbf{k}^2} + \frac{1}{\mathbf{k}^2}\right] + (\phi_t - 2n)\delta_{ml}. \qquad (11.185)$$

We consider next the *static* current-current susceptibility

$$\begin{aligned}
\chi_{ml}(\omega = 0, \mathbf{k}) &= (4n_s - \phi_t)\frac{k_m k_l}{\mathbf{k}^2} + (\phi_t - 2n)\delta_{ml} \\
&= (\phi_t - 2n)\left(\delta_{ml} - \frac{k_m k_l}{\mathbf{k}^2}\right) + (4n_s - 2n)\frac{k_m k_l}{\mathbf{k}^2} \quad (11.186)
\end{aligned}$$

The f-sum rule states that the longitudinal part of χ_{ml} is equal to $2n$ (with the particle mass $m = 1/2$). This fact implies that

$$4n_s - 2n = 2n \qquad \Rightarrow \qquad n_s = n. \qquad (11.187)$$

As a consequence, from the definition itself of the superfluid part we get:

$$\phi_t - 2n \equiv 2n_n = 2(n - n_s) = 0 \qquad \Rightarrow \qquad \phi_t = 2n. \qquad (11.188)$$

So we have also recovered the well known result for the zero temperature Bose Gas that $n_s = n$.

Eventually we get the following final expression:

$$\chi_{lm}(k) = \frac{c^2 n k_l k_m/m}{\omega^2 + c^2 \mathbf{k}^2} \qquad (11.189)$$

$$\Gamma_{;lm}(k) = \frac{n}{m}\delta_{lm} \qquad (11.190)$$

$$\mathcal{G}_{;lm}(k) = -\frac{n}{m}\delta_{lm} - \frac{c^2 n k_l k_m/m}{\omega^2 + c^2 \mathbf{k}^2}, \qquad (11.191)$$

where we have restored the usual dimension by reintroducing the mass m. Note that the *log* terms disappear in these expressions, leaving a regular function for $\mathcal{G}_{;\mu\nu}$ at this order in $k \to 0$.

We turn now our attention to the current-density response function. We begin with the following vertex functions [see (9.89)]:

$$\Gamma_{;0l}(p) = i k_l \omega \mathcal{P}_{;0v}. \qquad (11.192)$$

This means that for our purposes this vertex function can be treated as if it were equal to zero. Moreover by using the result obtained above for $\Gamma_{1;l}$ we are left with the following expression:

$$
\begin{aligned}
\mathcal{G}_{;0l}(p) &= -\Gamma_{1;0}(-p)\mathcal{G}_{12}(p)\Gamma_{2;l}(p) - \Gamma_{2;0}(-p)\mathcal{G}_{22}(p)\Gamma_{2;l}(p) \\
&= \frac{\mathbf{k}_l i\,\omega\,n_s/m}{\omega^2 + c^2\mathbf{k}^2}\,,
\end{aligned}
\tag{11.193}
$$

where again we have restored the usual dimensions.

11.4 Condensate Density

As a final check and a first implementation of the theory developed in the previous Chapters in this Section we use the RG machinery to investigate the appearance of corrections to the expression of the depletion of the condensate given by Bogolubov. We will verify that the leading order expression for the condensate density depends only on the invariants of the RG flow, thus coinciding with the mean field Bogolubov solution at the lowest order in the interaction.

1141   Bogulubov Condensate Density

We consider the Bogolubov expression for the condensate density [79]:

$$
n' \equiv n - n_0 = \frac{-1}{\beta\Omega}\sum_q \mathcal{G}(q)e^{i\omega 0^+}\,,
\tag{11.194}
$$

where

$$
\mathcal{G}(q) = -\frac{i\omega_\nu + \mathbf{q}^2 + \mu}{\omega_\nu^2 + E_{\mathbf{q}}^2}\,,
\tag{11.195}
$$

and

$$
E_{\mathbf{q}}^2 = 2\mu\mathbf{q}^2 + \mathbf{q}^4\,.
\tag{11.196}
$$

In this way for $T = 0$ we have:

$$
n' = \frac{1}{2}K_d\mu^{d/2}\int_0^\infty x^{d-1}dx\left\{\frac{x^2+1}{(x^4+2x^2)^{1/2}} - 1\right\}.
\tag{11.197}
$$

So the depletion depends only on the sound velocity $c = (2\mu)^{1/2}$. The contributions of the linear part of the spectrum ($x < 1$) and of the quadratic part ($x > 1$) are of the same order of magnitude.

1142 RG Condensate Density

Now we ask whether the RG result for the correlation functions can lead to correction to the well known-Bogolubov expression for the depletion of the condensate. We must remember that the Bogolubov approach is correct for the weak coupling limit, and in this limit the appearance of the divergences is limited to a very narrow region of the spectrum [87]. So in any case the correction coming from the correct RG treatment would be very small. In a more general case of strong interaction the singular region can extends over the whole linear part of the spectrum. In this case corrections could *a priori* be large. To inquire on the presence of these corrections we consider the expression for the density given by (11.194) and (11.161).

The depletion becomes:

$$n' = -\frac{1}{\beta\Omega} \sum_p [\mathcal{G}_{11}(p) + \mathcal{G}_{22}(p) - 2i\mathcal{G}_{12}(p)] \tag{11.198}$$

So we obtain for n' at $T = 0$:

$$n' = \frac{1}{\Omega} \sum_{\mathbf{k}} \frac{1}{v_{11}u_{22} + w_{12}^2} \frac{v_{11} + (z_{22} + z_{11})\mathbf{k}^2 - 2w_{12}E_{\mathbf{k}} - u_{22}E_{\mathbf{k}}^2}{2E_{\mathbf{k}}}, \tag{11.199}$$

where now:

$$E_{\mathbf{k}}^2 = \frac{v_{11}z_{22} + z_{11}z_{22}\mathbf{k}^2}{v_{11}u_{22} + w_{12}^2}\mathbf{k}^2 \to c^2\mathbf{k}^2. \tag{11.200}$$

These expressions are valid in the asymptotic region, where z_{11} is zero exponentially. So we can eliminate it soon. We begin by considering the lowest-order term, *i.e.*, we retain only v_{11} in the numerator:

$$n' \sim \frac{1}{\Omega} \sum_{\mathbf{k}} \frac{v_{11}}{v_{11}u_{22} + w_{12}^2} \frac{1}{2ck} = \frac{1}{\Omega} \sum_{\mathbf{k}} \frac{c}{2z_{22}k} = \frac{c}{z_{22}} \frac{K_d}{2} \int_0^c k^{d-2}dk \sim \frac{c^d}{z_{22}} \tag{11.201}$$

where

$$c^2 = \frac{v_{11}z_{22}}{v_{11}u_{22} + w_{12}^2}. \tag{11.202}$$

Recall that $z_{22} \sim n_s/n_0$, so the leading IR contribution to the depletion *depends only on the invariants of the RG flow.*

To see that this term is indeed the leading one, we can collect the v_{11} term in the numerator of (11.199), so to obtain the following expression:

$$v_{11} \left[1 - \frac{2w_{12}c}{v_{11}}k + \frac{z_{22} - u_{22}c^2}{v_{11}}k^2 \right]. \tag{11.203}$$

Now we know that v_{11} goes to zero, so it could seem that the contribution of the other terms becomes increasingly important. But this is not correct because, if we call Λ the upper cut-off of the IR theory, then the condition

$$\Lambda \ll c \tag{11.204}$$

holds for the weak coupling (or low density) limit [87]. It is easy to check (11.204), consider (11.160), in the weak coupling $n \sim n_0 \sim n_s$ and $c \sim c_o$, while we can choose as a good normalization point c_o: $\kappa \sim c_o$ so the $\ln(k/\kappa)$ can be written in a simplified way:

$$\mathcal{G}_{11}(k) \sim \frac{c_o}{64\pi^2 n} \ln(k/c_o) \qquad (11.205)$$

a comparison with the mean field value of the longitudinal correlation function (cfr. (8.24)) $\mathcal{G}_{11}(k) \sim c_0^{-2}$ gives the following value for Λ:

$$\Lambda = c_0 \exp\{-64\pi^2 n/c_o^3\} = c_0 \exp\{-64\pi^2/(2n^{1/3}v)^{3/2}\} \qquad (11.206)$$

so the weak coupling condition (that in this case where the UV cut-off is essentially c_o reads $v \ll c_o$ or also $n^{1/3}v \ll 1$) restrict the IR region in a very narrow region of the spectrum. From this fact follows that $k < \Lambda \ll c$.

In the weak coupling limit $\Lambda \to 0$ exponentially and the contribution of the singular terms remains always much smaller than the leading one:

$$(k/c_o)^n \ln(k/c_o) \ll 1 \qquad (11.207)$$

for $k < \Lambda \ll c_o$ is satisfied. So the low density expression for the depletion of the condensate found within the Bogolubov approach is indeed correct at the leading order.

Chapter 12

Conclusions of Part II

In Part II of the Thesis we have exploited the symmetries associated with the stability of the superfluid phase to solve the long-standing problem of interacting bosons in presence of a condensate at zero temperature. The Ward Identities implementing these symmetries have posed strong conditions on the renormalization required to heal the singularities of perturbation theory, and allowed to express the renormalization-group equations in terms of a limited set of physical quantities. The renormalized theory gives: For $d > 3$ the Bogolubov quasiparticle as an exact result (no divergences appear in the perturbation theory); for $1 < d < 3$ a non trivial solution with the exact exponent for the singular longitudinal correlation function and the exact IR behavior for the other propagators. $d = 3$ represents the upper critical dimension for which the longitudinal correlation function has a logarithmic singularity. In all the cases the low lying excitations turns out to be phonons.

In this way we have provided a consistent picture of the IR behavior of zero temperature interacting Bosons, in which (i) the vanishing of the anomalous self energy, (ii) the coincidence of the macroscopic with the microscopic sound velocity, and (iii) a non singular renormalized perturbation theory are found at the same time and with the same formalism.

Within this scheme, concrete calculations can be performed in terms of renormalized coupling, for instance the low density Bose gas result corresponds to the one-loop calculation presented above. It is worth noting again that higher order calculations will not spoil the result obtained for the IR behavior for the propagators and for the response functions, but will provide explicit ·expressions for the coefficients (for instance the compressibility) in terms of the interparticle potential.

The results obtained in this Part open interesting perspectives of research. First of all a comparison with the work of Popov within the hydrodynamical action in terms of the phase and modulus of the field should be investigated. This formalism allow a simple treatment of the IR divergences because implement directly the Gauge symmetry in the way in which the fields are written;

as a counterpart the perturbation theory becomes much more complicate and, in particular, it can't describe at the same time the UV and the IR behavior of the system. In fact, the cut-off k_o between "fast" and "slow" modes introduced by Popov is similar to the Wilson RG cut-off, but its expected disappearance from any physical quantities, contrary to the RG approach, is not easily controlled within the theory. To our purposes it can be interesting to use the RG to study the Popov action and make a comparison between the resulting fixed point actions. Part of this program has been already performed.

Extension of these results at finite temperature appears at this point possible and of particular interest. The study of the critical crossover from $T = 0$ to the critical temperature has been already object of study of Ref. [104] and references quoted therein, but we believe that still many questions are unanswered. In particular the interplay between the zeroth, first, and second sound as soon as the system is at finite temperature is still source of controversial in literature (see for instance Ref. [91]).

The results obtained at zero temperature suggest that the use of the renormalization group in this context can contribute to clear some of these oldstanding questions.

A RG approach to eliminate the UV and IR divergences in the theory can be set up, in this way it could be possible to follow the evolution of the correlation functions from the free particle to the phase fluctuations within the same formalism. Also along this line work has been performed and a complete account will be given in future publications. Some papers on the UV renormalization appeared recently [127,128] due to the interest for the low density Bose gas raised by the discovery of new Bose condensed systems [78].

General Conclusions

In this Thesis we faced two apparently different problems: the crossover from BCS to BEC superconductivity, and the infrared behavior of interacting bosons. We refer to the conclusions of the two Parts for a comment on the single topics, here we want only to give a final comment on the unity of the work done in this Thesis. In fact, the need of the study performed in Part II to approach the crossover problem of Part I can be now understood better. Not casually, in Part I we had to perform calculations only at the level of one-loop for Fermions, while in Part II we investigated the whole perturbation theory for bosons. The infrared divergences that plague the Bose theory appear in fact in the Fermi theory at the *next* order with respect the same theory for bosons. The singular behavior of the two systems has exactly the same origin, *i.e.*, the Goldstone mode singularity. It originates from the broken gauge symmetry that is the unifying feature of the two phenomena.

At this point, using the results obtained for the Bose system as a guide, it seems feasible a similar treatment for the theory of superfluid fermions and a general better understanding of the associated problems.

It is also worth saying that during the final stages of this PhD work the interest on the crossover problem has been greatly renewed by new experimental results [129–131]. There is in fact evidence of a pseudo gap in the single particle spectrum above the critical temperature. This gap disappears only for temperature much greater that T_c for the underdoped materials and can be interpreted as a signature of "preformed" pairs above T_c. If this is the case, the system appears to be exactly in the crossover region, but still with a well defined Fermi surface (in our picture of the crossover this would imply $k_F \xi_{pair} \sim 5 - 10$). This region is the most difficult to treat and many recent papers are devoted to improve our knowledge on this strongly correlated system [132–137]. We believe that a deeper understanding of the Bose system can help greatly the comprehension of the crossover problem [138], and can give a guide in the way it can be attacked.

Acknowledgments

It is a really pleasure for me to thank Giancarlo, who followed my work since the Diploma Thesis, and has become for me a guide not only in physics. His strong passion for the physics together with the aim of understanding deeply each problem has been and will be always a stimulating example for me. I can't forget to thank also his wife Claudia for her kindness and also for her comprehension for all the times that the discussions with Giancarlo at University of Rome finished too late, and sometimes continued at their home where I have been guest many times.

I express a special thank to Prof. Claudio Castellani and Prof. Carlo Di Castro who formulated with us the research project of the second part of this Thesis and with whom we developed it. I believe that during the long discussions in Carlo's office with Giancarlo and Claudio I really learned a beautiful method of doing physics.

I am indebted to Prof. G. Franco Bassani for his confidence in my research project and continuous support during the course of this work.

I also acknowledge Prof. M.P. Tosi and Prof. F. Beltram for interesting discussions.

For having encouraged this research and valuable discussions in Rome and Grenoble I acknowledge Prof. P. Nozières.

I gratefully acknowledge Europa Metalli LMI S.p.A. that financed the Ph.D. program at the Scuola Normale Superiore that has made possible this work. The italian INFM is acknowledged for partial financial support.

It is not possible to make a complete list of all the people of "Via della Faggiola" that shared with me these years: P. Giannozzi, F. Buda, G. La Rocca, S. Moroni, R. Sholtz E. Arrigoni, A. Morpurgo, V. Tozzini, S. Conti, M.L. Chiofalo, A. Tredicucci, V. Pellegrini, S. De Franceschi, S. De Gironcoli, R. Atanasov, A. Bifone, F. Tassone and many others. I am glad to see that the *porco rampante*'s crew is growing every year.

My "zia Peppina" in Rome and his family: Marina and Federico, Mauro and Vanda are all aknoledged for their ospitality and the colorful Roman spirit with whom they received me.

I want also to thank my mother, my brother and particularly Cristina for continuous support and encouragements during the difficult moments.

Bibliography

[1] "Evolution from BCS superconductivity to Bose condensation: Role of the parameter $k_F\xi$", F. Pistolesi and G.C. Strinati, Phys. Rev. B **49**, 6356 (1994)

[2] "Evolution from BCS Superconductivity to Bose Condensation: Universal Behaviour with the Model of Nozières and Schmitt-Rink" F.Pistolesi and G.C. Strinati, in **BEC93** edited by D. Snoke, A. Griffin, and S. Stringari (Cambridge Univ. Press (1994)).

[3] "Revisiting the Nozières and Schmitt-Rink Approach for the Evolution from BCS Superconductivity to Bose Condensation" F. Pistolesi and G.C. Strinati, Physica C **235-240** 2359 (1994).

[4] "Evolution from BCS Superconductivity to Bose Condensation: Calculation of the Zero-Temperature Phase Coherence Length", F. Pistolesi and G.C. Strinati, Phys. Rev. B. **53**, 15168 (1996).

[5] "Infrared Behavior of Interacting Bosons at Zero Temperature" C. Castellani, C. Di Castro, F. Pistolesi, and G.C. Strinati, cond-mat/9604076, Phys. Rev. Lett. **78**, 1612 (1998).

[6] "Renormalization Group Approach to the Infrared Behavior of the Zero-Temperature Bose System" F. Pistolesi, C. Castellani, C. Di Castro, and G.C. Strinati, in preparation.

[7] M. Randeria, Ji-Min Duan, and Lih-Yir Shieh, Phys. Rev. Lett. **62**, 981 (1989).

[8] M. Randeria, Ji-Min Duan, and Lih-Yir Shieh, Phys. Rev. B **41**, 327 (1990).

[9] L. Belkhir and M. Randeria, Phys. Rev. B **45**, 5087 (1992).

[10] J. O. Sofo, C. A. Balseiro, and H. E. Castillo, Phys. Rev. B **45**, 9860 (1992).

[11] T. Kostyrko and R. Micnas, Phys. Rev. B **46**, 11025 (1992).

[12] A. S. Alexandrov and S. G. Rubin, Phys. Rev. B **47**, 5141 (1993).

[13] C. A. R. Sá de Melo, M. Randeria, and J. R. Engelbrecht, Phys. Rev. Lett. **71**, 3202 (1993).

[14] L. Belkhir and M. Randeria, Phys. Rev. B **49**, 6829 (1994).

[15] Before the discovery of high temperature superconductors introduced a Boson-Fermion model to treat the many polaron system: J. Ranninger and S. Robaszkiewicz, Physica B **135**, 468 (1985).

[16] R. Friedberg and T.D. Lee, Phys. Rev. B **40**, 6745 (1989).

[17] S. Schmitt-Rink, C. M. Varma and A. E. Ruckenstein, Phys. Rev. Lett. **63** 445 (1989).

[18] A. Moreo and D. J. Scalapino, Phys. Rev. Lett. **66**, 946 (1991).

[19] J.G. Bednortz and K. A. Muller Z. Phys. B **64**, 189 (1986).

[20] Y. J. Uemura *et al.*, Phys. Rev. Lett. **66**, 2665 (1991); Nature **352**, 605 (1991).

[21] The crossover of ξ_{phase} from BCS to BE has been shortly discussed in Ref. 13, expanding the action of the functional integral near the critical temperature where the gap parameter is small. We have considered instead the broken-symmetry state at zero temperature directly. For this reason, our analysis should be more reliable and exhaustive.

[22] P. Nozières and S. Schmitt-Rink, J. Low. Temp. Phys. **59**, 195 (1985).

[23] L. V. Keldish and Y. V. Kopaev, Sov. Phys. Solid State **6**, 2219 (1965).

[24] L. V. Keldish and A. N. Kozlov, Sov. Phys. JETP **27**, 521 (1968).

[25] D. M. Eagles, Phys. Rev. **186**, 456 (1969).

[26] A. J. Leggett, in *Modern Trends in the Theory of Condensed Matter*, A. Pekalski and J. Przystawa, Eds., Lecture Notes in Physics Vol. 115 (Springer-Verlag, Berlin, 1980), p. 13.

[27] J. O. Sofo and C. A. Balseiro, Phys. Rev. B **45**, 8197 (1992), and references quoted therein.

[28] G. Baym and L. P. Kadanoff, Phys. Rev. **124** 287 (1961). G. Baym, Phys. Rev. **127**, 1391 (1962).

[29] R. Haussmann, Z. Phys. B **91**, 291 (1993); Phys. Rev. B **49**, 12975 (1994).

[30] Cf. , e.g., G. Rickayzen, *Theory of Superconductivity* (Interscience, N. Y., 1965), Chap. 4.

[31] P. C. Hohenberg and P. C. Martin, Ann. Phys. **34**, 291 (1965), and references quoted therein.

[32] As a consequence, satisfying the conservation laws for the *effective* bosonic system resulting from a superconducting fermionic system in the extreme strong-coupling limit poses no problem (cf. Ref. 29).

[33] An extensive literature exists on the *dilute* Bose gas. For recent reviews, see: P. Nozières and D. Pines, *The Theory of Quantum Liquids* (Addison-Wesley, Redwood City, Cal., 1990), Vol. II; A. Griffin, *Excitations in a Bose-Condensed Liquid* (Cambridge Univ. Press, Cambridge, 1993).

[34] D. E. Soper, Phys. Rev. D **18**, 4590 (1978).

[35] V. N. Popov, *Functional Integrals in Quantum Field Theory and Statistical Physics* (Riedel, Dordrecht, 1983); V. N. Popov, *Functional Integrals and Collective Excitations* (Cambridge Univ. Press, Cambridge, 1987).

[36] Although a functional-integral formulation has been already used in Ref. 13 to recover the NSR crossover from BCS to Bose superconductivity, apparently the need of going beyond Gaussian fluctuations to be consistent with the loop expansion has not been pointed out in that paper.

[37] S. V. Traven, Phys. Rev. Lett. **73**, 3451 (1994).

[38] A detailed proof of this statement will be given elsewhere.

[39] F. Pistolesi and G. C. Strinati, Phys. Rev. B **49**, 6356 (1994).

[40] The condition $k_F \xi_{pair} \simeq 10$ has been identified as the beginning of the crossover from BCS to BE also by an approach based on Eliashberg theory for superconductivity (G. Varelogiannis and L. Pietronero, unpublished).

[41] In a recent paper by M. Casas *et al.* (Phys. Rev. B **50**, 15945 (1994)) the coherence length for two-electron correlation (namely, ξ_{pair} of the present Thesis) was calculated at the mean-field level for a momentum-independent gap energy and compared with available experimental data on cuprate superconductors. Since the data refer to ξ_{phase} instead, a comparison of this kind is meaningful as far as $\xi_{phase} \simeq \xi_{pair}$, which we show in the present Thesis to hold provided $k_F \xi_{pair} \gtrsim 10$.

[42] Y.J. Uemura *et al.* Phys. Rev. Lett. **66**, 2665 (1991); Nature **352**, 605 (1991). Ch. Niedermayer *et al.*, Phys. Rev. Lett. **71**, 1764 (1993).

[43] Cf., e.g., J.R. Schrieffer, *Theory of Superconductivity* (Benjamin, New York, 1964), Chapter 2.

[44] Reduction to an effective fermionic attractive potential may not be possible whenever dynamical (retardation) effects are important. For recent developments see L. Pietronero and S. Strässler, Europhys. Lett. **18**, 627 (1992).

[45] These two criteria were borne out in the original BCS suggestion to express their theoretical results in the form of *ratios* of experimentally accessible quantities, which are independent of the interaction potential and of the single-particle density of states.

[46] As in the BCS approach, we rely on a description of the normal state in terms of quasiparticles.

[47] We have got rid at the outset of the Hartree-Fock-like terms by setting $V_{k,k} = 0$ for the diagonal components. This choice will by no means invalidate our results, since it turns out that these terms are irrelevant in the parameter region of physical interest.

[48] NSR (Ref. [22]) state instead that $r_0 \sim k_0^{-1}$ for $G \to \infty$. Since bosonization can be achieved only when $r_0^3 \ll n^{-1}$, NSR are able to follow the evolution from BCS to BE condensation as a function of G *only* in the "dilute limit" $n/k_0^3 \ll 1$ for the reduced (three-dimensional) density, that is, for given density n only when $k_0 \gg k_F$. This limitation has prevented NSR from connecting the two physical limits (BCS and BE) irrespective of k_0. Our finding that $r_0 \sim k_0^{-1} G^{-1/2}$, on the other hand, enables us to satisfy the bosonization condition $(n/k_0^3)/G^{3/2} \ll 1$ even in the "dense limit" $n/k_0^3 \gg 1$, provided G is large enough.

[49] Probably the most suited quantity to follow the evolution from BCS superconductivity to BE condensation is the chemical potential μ, which almost coincides with the Fermi energy in weak-coupling limit and reduces to (half of) the lowest eigenvalue of the associated two-body problem in the strong coupling limit. Furthermore, μ can also be object of direct measurements: cf. G. Rietveld, N.Y. Chen, and D. van der Marel, Phys. Rev. Lett. **69**, 2578 (1992), and references quoted therein.

[50] A physical quantity directly related to xi is the (average) numbers of electrons $\mathcal{N}_d$ within a correlation volume in d dimensions. When $d = 3$ one gets $\mathcal{N}_3 = (k_F\xi)^3/18\pi$, while when $d = 2$ one gets $\mathcal{N}_2 = (k_F\xi)^2/8$. Since in the Uemura plot three-dimensional as well as (quasi) two-dimensional systems are treated on the same footing, we prefer to use simply ξ as the variable replacing G.

[51] We have verified that the universal behavior shown in Fig. 2.1 for $\xi \gtrsim 2\pi$ and $\xi \lesssim \pi^{-1}$ is *independent* of the choice of the single-particle dispersion relation $\epsilon_{\mathbf{k}}$ and of the shape of the interaction potential embodied by the function $w_{\mathbf{k}}$ provided obviously that these functions do not behave in a pathological way in $\mathbf{k}$ space.

[52] The difference between the numerical prefactors of Eqs.(2.8) and (2.9) is due to the difference between our definition (2.7) for ξ and the Pippard coherence length ξ_0. We have also verified that Eq.(2.9) holds for $\xi \gtrsim 2\pi$ both in three and in two dimensions. This is a requisite to apply Eq.(2.9) to the Uemura plot, since the data there reported pertain to fully three-dimensional as well as to (quasi) two-dimensional systems.

[53] It is interesting to apply Eq.(2.9) also to superfluid ^{3}He for which $k_F = 0.78\text{Å}^{-1}$, $m^*/m = 2.76$ (where m is the bare mass of the ^{3}He atom), and $\xi \approx 200\text{Å}$. Equation (2.9) then provides for T_c the value 4.5 mK, which compares reasonably well with the experimental value (2.6 mK) if we consider the fact that Eq. (2.9) applies to s-wave pairing while the pairing in ^{3}He is known to be p-wave. It is thus essentially the large value of ξ (≈ 160) which accounts for the three orders of magnitude difference between the superfluid temperature of ^{3}He and ^{4}He (we are indebted to C. Di Castro for this remark). One should also mention in this context that common features between ^{3}He and heavy-fermion systems have already been pointed out by R. Tournier *et al.*, J. Magnetism and Magn. Mat. **76-77**, 552 (1988) (see especially their Fig. 6). (We are indebted to J. Ranninger for bringing this reference to our attention).

[54] When comparing the values of the coherence length ξ obtained by our definition (2.7) with the experimental values, one should be aware of the fact that different numerical factors (of order unity) plague alternative definitions of the (zero-temperature) coherence length. The relation between our ξ and the experimentally determined ξ_{exp} can be determined by relating ξ_{exp} to the Pippard coherence length ξ_0, in the limit when a microscopic derivation of the Ginzburg-Landau equation from BCS is justified. From the clean-limit expression $\xi_{exp} = 0.87\xi_0$ [cf., E. Helfand and N.R. Werthamer, Phys. Rev. **147**, 288 (1966), and references quoted therein] we then obtain $\xi \approx 1.25\xi_{exp}$, which we assume to apply for *all* materials reported in the Uemura plot.

[55] See, e. g., D. Forster, *Hydrodynamic Fluctuations, Broken Symmetry, and Correlation Functions* (Benjamin, Mass., 1975), Chaps. 10 and 12; see also P. B. Weichman, Phys. Rev. B **38**, 8739 (1988).

[56] We set $\hbar = 1$ throughout.

[57] A Hamiltonian similar to (3.12) results from the so-called negative-U Hubbard model in a lattice, whereby $V(\mathbf{k}, \mathbf{k}')$ is replaced by the constant U, $\xi_{\mathbf{k}}$ by the band dispersion relation, and the wave vectors are restricted to the Brillowin zone. In Appendix D we consider this model Hamiltonian at the mean-field level.

[58] The Hamiltonian (3.12) is, in general, invariant under a global but not a local gauge transformation. Local gauge invariance is suitably recovered whenever the interaction potential $V(\mathbf{k}, \mathbf{k}')$ depends weakly on $\mathbf{k}$ and $\mathbf{k}'$, i.e., in the limit $k_0 \to \infty$ for the choices (3.13) and (3.14).

[59] For a recent pedagogical review on how path integrals can be written for fermions, see R. Shankar, Rev. Mod. Phys. **66**, 129 (1994), Sect. III, and references quoted therein.

[60] The separable potential (3.13) has led us to represent the interaction part of the Hamiltonian as a bilinear form in the operators (3.19). When $w(\mathbf{k})$ depends weakly on $\mathbf{k}$, an alternative decoupling in terms of the density fluctuation operators $\rho_\sigma(\mathbf{q}, \tau) = \sum_{\mathbf{k}} \bar{c}_\sigma(\mathbf{k}+\mathbf{q}, \tau) c_\sigma(\mathbf{k}, \tau)$ is also possible. As discussed in Ref. 11, linear responses for the two operators $\rho_\sigma(\mathbf{q})$ and $\mathcal{B}(\mathbf{q})$ are, quite generally, coupled. In the following, we shall not consider this coupling at the Gaussian level since it is not essential for our purposes.

[61] To be more precise, the time label that specifies $\dot{\bar{\mathcal{B}}}(\mathbf{q})$ should be augmented by a positive infinitesimal ($\tau \to \tau + \delta$ with $\delta = 0^+$) on both sides of Eq. (3.20). This implies that the Fourier transformed variables have to be interpreted (in the mixed fermionic-bosonic terms only) as $b(\mathbf{q}, \omega_\nu) \to b(\mathbf{q}, \omega_\nu) e^{i\omega_\nu \delta}$ and $b^*(\mathbf{q}, \omega_\nu) \to b^*(\mathbf{q}, \omega_\nu)$.

[62] In other words, the identity $f(\lambda) = 0$ entails the vanishing of every coefficient f_n of the expansion $f(\lambda) = \sum_{n=0}^{\infty} f_n \lambda^n$.

[63] The standard BCS expression for the grand-canonical free energy at temperature β^{-1}, namely, $F_0^{(BCS)} = -\frac{|\Delta_0|^2}{V} - \frac{2}{\beta} \sum_{\mathbf{k}} \ln\left(1 + e^{-\beta E_{\mathbf{k}}}\right) + \sum_{\mathbf{k}} (\xi_{\mathbf{k}} - E_{\mathbf{k}})$, results from Eq. (3.37) *provided* the sum over the fermionic frequencies ω_s (which is implicit in the trace therein) is evaluated carefully by the correct time discretization procedure of the fermionic functional integral.

[64] In Eqs. (3.39) and (3.40) the symmetry properties of $A(q)$ and $B(q)$ when $q \to -q$ hold provided $w(-\mathbf{k}) = w(\mathbf{k})$.

[65] Hartree-Fock terms do not appear in the expression of F_0 owing to our mean-field decoupling; they appear instead in the expression of F_1, as one can readily verify for the "normal" ($\Delta_0 = 0$) phase by expanding F_1 in powers of V.

[66] E. Arrigoni, C. Castellani, M. Grilli, R. Raimondi, and G. C. Strinati, Phys. Reports **241**, 291 (1994).

[67] Besides the diagrams depicted in Fig. 3.3, the "exchange" contribution (3.50) contains also terms with the structure of Eq. (3.55), where the fermionic bubbles (A and B) are calculated with the "dressed" version of the single-particle fermionic Green's functions that include corrections of order λ (cf. the first term on the right-hand side of Eq. (3.52)). Despite this replacement, this contribution is expected to maintain the "short-range" character of the mean-field expression (3.55) and is accordingly neglected in the following.

[68] More precisely, the characteristic spatial "range" associated with a (non-negative) function $F(\mathbf{r})$, whose Fourier transform $F(\mathbf{q})$ is well defined, can be determined by

$$\left\langle r^2 \right\rangle \equiv \frac{\int d\mathbf{r}\ F(\mathbf{r})\mathbf{r}^2}{\int d\mathbf{r}\ F(\mathbf{r})} = \frac{-\nabla_\mathbf{q}^2 F(\mathbf{q})\big|_{q=0}}{F(\mathbf{q}=0)}.$$

This justifies retaining the quadratic expansion (3.64) only for the calculation of ξ_{phase}.

[69] The BCS limit is identified by the condition $\Delta_{k_F} \ll \mu$ with $\mu > 0$. In this limit one can assume, in addition, that $\Delta_{k_F} \ll k_0^2/2m$ for any given k_0 (k_0 being the characteristic wave vector of the interaction potential - cf. Eq. (3.14)), provided one takes the strength V sufficiently small (cf. Eq. (3.13)). With these two conditions, one verifies that the nonvanishing contributions to expression (3.67) (other than the contribution (3.69)) are smaller than the dominant contribution (3.69) by factors $\left(\frac{\Delta_{k_F}}{\mu}\right)\left(\frac{\Delta_{k_F}}{k_0^2/2m}\right)$ or $\left(\frac{\Delta_{k_F}}{k_0^2/2m}\right)^2$. For the approximations used to derive the right-hand side of Eqs. (3.69) and (3.70), see also Appendix C.

[70] The BE limit is achieved when the condition $\Delta_0 \ll |\mu|$ with $\mu < 0$ is satisfied. In this limit, however, the characteristic wave vector k_0 of the interaction plays an important role (in contrast with the BCS limit discussed in Ref. 69, for which the value of k_0 is irrelevant). The point is that increasing the value $|V|$ of the interaction strength makes $|\mu|$ increase accordingly, and thus $|\mu|$ becomes eventually larger than $k_0^2/2m$ for any initial choice of k_0. Expression (3.67) for the coefficient b, in turn, depends crucially on the ratio $|\mu|/(k_0^2/2m)$ and different values are obtained depending on which of the two conditions ($|\mu| \ll k_0^2/2m$ or $|\mu| \gg k_0^2/2m$) is satisfied. We have thus to impose some *restrictions* on the bosonization procedure according to the following scheme. For given k_0, $|V|$ is increased

to reach the bosonization condition $\Delta_0 \ll |\mu|$ with $\mu < 0$, paying attention that $|\mu|$ remains much smaller than $k_0^2/2m$. If this is the case, there is no need to increase $|V|$ any further and the bosonization condition has *effectively* been achieved. Otherwise, if $|\mu|$ needs to become comparable to or even larger than $k_0^2/2m$ to reach the bosonization condition $\Delta_0 \ll |\mu|$ with $\mu < 0$, the calculation might produce inconsistent results, such as negative values for the coefficient b. We shall thus impose the *further condition* $|\mu| \ll k_0^2/2m$ on the BE limit (which is equivalent to a low-density condition for the Bose gas). With this condition, one verifies that the terms of the expression (3.67) (other than the contribution (3.73)) are smaller than the dominant contribution (3.73) by factors $\left(\frac{\Delta_0}{|\mu|}\right)$ and/or $\left(\frac{\Delta_0}{k_0^2/2m}\right)$ and their powers.

[71] It is interesting to derive from Eqs. (3.81) and (3.82) the limiting value of ξ_{phase}^{BE} for $k_0 \to \infty$, whereby the interaction potential (in real space) reduces to a "contact" potential. Implementing this limit requires one to keep ξ_{pair}^{BE} constant, in such a way that $|\mu_0|$ remains also constant and $c \to 0$ in Eq. (3.82). One then obtains $(k_F \xi_{phase}^{BE})^2 = (3\pi/16)\sqrt{|\mu_0|/\epsilon_F}$ with $\epsilon_F = k_F^2/2m$. Alternatively, relating $|\mu_0|$ to ξ_{pair}^{BE} (cf. Appendix C) one rewrites $(k_F \xi_{phase}^{BE})^2 = (3\pi/16\sqrt{2})(k_F \xi_{pair}^{BE})^{-1}$, which coincides (apart possibly for a numerical factor of order unity) with the result reported in Ref. 13 in terms of the scattering amplitude for the two-fermion problem.

[72] A related problem has been addressed in Ref. 29 using conventional diagrammatic techniques, where the two-fermion correlation function has been shown to reduce to the Bogolubov propagator in the strong-coupling limit. The functional-integral approach enables us to go further and study higher-order effects such as multiple-boson interactions.

[73] In the broken-symmetry state at zero temperature, the quadratic action (3.38) can be expanded in series of the small parameter Δ_0^2/μ^2 in the BE limit [cf. Eq. (4.24)]. In the absence of loops for the bosonic propagators, retaining the lowest significant order in Δ_0^2/μ^2 is equivalent to keeping $v_2(0)$ only out of the set $v_n(0)$ given by Eq. (4.29). In bosonic language, in fact, increasing n by one unit introduces in the self-energy two additional *condensate* lines proportional to $|\langle b'(q=0)\rangle|^2 \sim \beta\Omega|\mu|^{d/2}(\Delta_0/\mu)^2$ [cf. Eqs. (3.47), (4.13), and (4.18)].

[74] The bosonization criterion (4.22), which is exploited to obtain the effective bosonic action from the original fermionic action, ensures also that the resulting bosonic system is *dilute*. From the analysis of Appendix B one, in fact, obtains in three dimensions that the bound-state radius r_0 coincides with the scattering amplitude a_s for the two-fermion problem (apart from

a numerical factor of order unity). Condition (4.22) is thus equivalent to the standard criterion $a_s^B n_B^{1/3} \ll 1$ for a dilute Bose gas, where $a_s^B = 2a_s$ and $n_B = n/2$.

[75] Cf., e. g., P. Nozières, in *Bose-Einstein Condensation* (Cambridge Univ. Press, Cambridge, 1995), A. Griffin, D. W. Snoke, and S. Stringari, Eds., p. 15.

[76] The coupling between longitudinal and transverse fluctuations has been formulated via the general principle of "conservation of the modulus" by A. Z. Patashinskii and V. L. Pokrovskii, Zh. Eksp. Teor. Fiz. **64**, 1445 (1973) [Sov. Phys. JEPT **37**, 733 (1973)].

[77] Yu. A. Nepomnyashchii and A. A. Nepomnyashchii, Zh. Eksp. Teor. Fiz. **75**, 976 (1978) [Sov. Phys. JEPT **48**, 493 (1978)].

[78] M. H. Anderson *et al.*, Science **269**, 198 (1995).

[79] N. Bogolubov, J. Phys. **11**, 23 (1947).

[80] S. Beliaev, Sov. Phys. JETP **34**, 289 (1958).

[81] N. Hugenholtz and D. Pines, Phis. Rev. **116**, 489 (1959).

[82] J. Gavoret and P. Nozieres, Ann. Phys. (N.Y.) **28**, 349 (1964).

[83] K. Huang and A. Klein, Ann. Phys. (N.Y.) **30**, 203 (1964).

[84] F. De Pasquale, G. Jona-Lasinio, and E. Tabet, Ann. Phys. (N.Y.) **33**, 381 (1965).

[85] P. C. Hohemberg and P. C. Martin, Ann. Phys. **34**, 291 (1965).

[86] A. Nepomnyashchy and Y. Nepomnyashchy, Sov. Phys. JETP Lett. **21**, 1 (1975). See also Ref. 77. Y. Nepomnyashchy, Phys. Rev. B **46**, 6611 (1992).

[87] V. Popov and A. V. Seredniakov, Sov. Phys. JETP **50**, 193 (1979).

[88] P. B. Weichman, Phys. Rev. B **38**, 8739 (1988).

[89] P. Nozières and D. Pines, *The Theory of Quantum Liquids* (Addison-Wesley, Redwood City, Cal., 1990), Vol. II;

[90] S. Giorgini, L. Pitaevskii, and S. Stringari, Phys. Rev. B **46**, 6374 (1992).

[91] A. Griffin, *Excitations in a Bose-Condensed Liquid* (Cambridge Univ. Press, Cambridge, 1993).

[92] G. Benfatto, in "Constructive Results in Field Theory, Statistical Mechanics, and Condensed Matter Physics", Palaiseau, July 25-27, (1994).

[93] A. Griffin, Phys. Rev. B, **53**, 9341 (1996).

[94] A. Z. Patašinskij and V. L. Pokrovskij, Sov. Phys.-JETP **37**, 733 (1974).

[95] Without loss of generality we assume that the order parameter is real.

[96] The factor 2 has been introduced in (8.19) to obtain the simple relation between M and $\mathcal{G}^o$. Note that the 2 comes from the fact that the fields $\tilde{\chi}(k)$ are not independent and when we perform the functional integration to calculate the propagator we should write them in terms only of the independent ones. This brings a factor 2 in the action, and so a factor $1/2$ in the Green's function.

[97] G. 't Hooft and M. Veltman, Nuclear Physics, B **44**, 189 (1972).

[98] D. J. Amit, *Field Theory, the Renormalization Group, and Critical Phenomena* (World Scientific, Singapore, 1984).

[99] Here we must remember the problems arising from the correct time ordering of the operators, that in Eq. (9.146) has been completely lost. To calculate the total density in Sec. 11.4 we will use the correct time ordered functions.

[100] G. Baym, in *Mathematical Method od Solid State and Superfluid Theory*, ed. R.C. Clarck (Oliver and Boyd, Edinburg, 1965) p. 121.

[101] M.N. Barber, J. Phys. A: Math. Gen. **10**, 1335 (1977).

[102] M. E. Fisher, M. N. Barber, D. Jasnow, Phys. Rev. B **8**, 1111, (1973).

[103] K. K. Singh, Phys. Rev. B **12**, 2819, (1975).

[104] P. B. Weichman, M. Rasolt, M. E. Fisher, and M. J. Stephen, Phys. Rev. B **33**, 4632 (1986), and references quoted therein.

[105] D. S. Fisher and P. C. Hohenberg, Phys. Rev. B **37**, 4936 (1988).

[106] D. J. Wallace and R. K. P. Zia, Phys. Rev. B **12**, 5340 (1975).

[107] J. Rudnick and D. R. Nelson, Phys. Rev. B **13**, 2208 (1976).

[108] D. R. Nelson, Phys. Rev. B **13**, 2222 (1976).

[109] J. F. Nicoll, T. S. Chang, and H. E. Stanley, Phys. Rev. B **13**, 1251 (1976).

[110] J. F. Nicoll and T. S. Chang, Phys. Rev. A **17**, 2083 (1978).

[111] L. Schäfer and H. Horner, Z. Physik B **29**, 251 (1978).

[112] A similar RG approach was used in Ref. [92], although limited to $d = 3$ and without a systematic use of WI.

[113] See, *e.g.*, C. Domb and M.S. Green, eds., *Phase Transitions and Critical Phenomena*, Vol. **VI** (Academic Press, N.Y., 1976).

[114] The vanishing of v_{11} and so of the anomalous self-energy is the analogue of the divergence of the longitudinal correlation function in spin-wave theory [94], whereby the presence of the Goldstone mode in the transverse correlation function drives a weaker singularity in the longitudinal one.

[115] For a similar problem with Fermi systems see, *e.g.*, W. Metnezer and C. Di Castro, Phys. Rev. B **47**, 16107 (1993).

[116] S. K. Ma and C. Woo, Phys. Rev. **159**, 165 (1967).

[117] A. Griffin and T. H. Cheung, Phys. Rev. A **7**, 2086 (1973).

[118] P. Szèpfalusy and I. Kondor, Ann. Phys. (N.Y.) **82**, 1 (1974).

[119] V. K. Wong and H. Gould, Phys. Rev. B **14**, 3961 (1976).

[120] V. K. Wong and H. Gould, Ann. Phys. (N.Y.) **83**, 252 (1974).

[121] T. Toyoda Ann. Phys. (N.Y.) **141**, 154 (1982).

[122] T. Toyoda Ann. Phys. (N.Y.) **147**,244 (1983).

[123] E. Talbot and A. Griffin, Ann. Phys. (N.Y.) **151**, 71 (1983).

[124] E. Talbot and A. Griffin, Phys. Rev. B **29**, 3952 (1984).

[125] H.R. Glyde and A. Griffin, Phys. Rev. Lett. **65**, 1454 (1984).

[126] The *ln* terms in the ψ-representation have been found Ref. [86] without identifying their coefficients. The expressions for the dilute gas were given in Ref. [88], while in Ref. [90] the coefficients were calculated using the Landau Hydrodynamic Hamiltonian assumptions.

[127] E. Braaten and A. Nieto, cond-mat/9609046, (unpublished).

[128] M. Bijlsma and H.T.C. Stoof, cond-mat/9607188, (unpublished).

[129] Ding H. *et al.*, Phys. Rev. Lett. **74**, 1533 (1996)

[130] Ding H. *et al.*, Nature, **382**, 51 (1996), and reference quoted therein.

[131] see also B. Batlog and V.J. Emery, bf 382, 20, (1996) for some comments on these new experimental results.

[132] J.R. Engelbrecht, M. Randeria, and C.A.R. Sà de Melo, (unpublished).

[133] J. Ranninger and J. M. Robin, cond-mat/9603084 Phys. Rev. B. **XX**, RXXXX, (1996), and references quoted therein.

[134] J. Maly, K. Levin, and D.Z. Liu, cond-mat/9609083.

[135] O. Tchernyshyov and Y.J. Uemura, Columbia University preprint CU-TP-768 (unpublished).

[136] O. Tchernyshyov, Columbia University preprint CU-TP-773 (unpublished).

[137] V.B. Geshkenbein, L.B. Ioffe, and A.I. Larkin, cond-mat/9609209.

[138] Difficulties and unsolved problems on a similar system (the Mott transition) are pointed out in P. Nozières, Ann. Phys. (Paris) C2 **20** 417 (1995).

Elenco delle Tesi di perfezionamento della Classe di Scienze
pubblicate dall'Anno Accademico 1992/93

HISAO FUJITA YASHIMA, *Equations de Navier-Stokes stochastiques non homogènes et applications*, 1992.

GIORGIO GAMBERINI, *The minimal supersymmetric standard model and its phenomenological implications*, 1993.

CHIARA DE FABRITIIS, *Actions of Holomorphic Maps on Spaces of Holomorphic Functions*, 1994.

CARLO PETRONIO, *Standard Spines and 3-Manifolds*, 1995.

MARCO MANETTI, *Degenerations of Algebraic Surfaces and Applications to Moduli Problems*, 1995.

ILARIA DAMIANI, *Untwisted Affine Quantum Algebras: the Highest Coefficient of* det H_η *and the Center at Odd Roots of 1*, 1995.

FABRIZIO CEI, *Search for Neutrinos from Stellar Gravitational Collapse with the MACRO Experiment at Gran Sasso*, 1995.

ALEXANDRE SHLAPUNOV, *Green's Integrals and Their Applications to Elliptic Systems*, 1996.

ROBERTO TAURASO, *Periodic Points for Expanding Maps and for Their Extensions*, 1996.

YURI BOZZI, *A study on the activity-dependent expression of neurotrophic factors in the rat visual system*, 1997.

MARIA LUISA CHIOFALO, *Screening effects in bipolaron theory and high-temperature superconductivity*, 1997.

DOMENICO M. CARLUCCI, *On Spin Glass Theory Beyond Mean Field*, 1998.

RENATA SCOGNAMILLO, *Principal G-bundles and abelian varieties: the Hitchin system*, 1998.

GIACOMO LENZI, *The MU-calculus and the Hierarchy Problem*, 1998.

GIORGIO ASCOLI, *Biochemical and spectroscopic characterization of CP20, a protein involved in synaptic plasticity mechanism*, 1998.

FABIO PISTOLESI, *Evolution from BCS Superconductivity to Bose-Einstein Condensation and Infrared Behavior of the Bosonic Limit*, 1998.

"CompoMat" Loc. Braccone, 02040 Configni (RI), Italy
Finito di stampare nel febbraio 1999